Interactive Guide

Grade K

ConnectED.mcgraw-hill.com

Copyright © 2014 McGraw-Hill Education

All rights reserved. No part of this publication may be reproduced or distributed in any form or by any means, or stored in a database or retrieval system, without the prior written consent of McGraw-Hill Education, including, but not limited to, network storage or transmission, or broadcast for distance learning.

STEM McGraw-Hill is committed to providing instructional materials in Science, Technology, Engineering, and Mathematics (STEM) that give all students a solid foundation, one that prepares them for college and careers in the 21st century.

Send all inquiries to:
McGraw-Hill Education
8787 Orion Place
Columbus, OH 43240

Selections from:
ISBN: 978-0-02-133102-4 *(Grade K Student Edition)*
MHID: 0-02-133102-2 *(Grade K Student Edition)*
ISBN: 978-0-02-131855-1 *(Grade K Teacher Edition)*
MHID: 0-02-131855-7 *(Grade K Teacher Edition)*

Printed in the United States of America.

Visual Kinesthetic Vocabulary® is a registered trademark of Dinah-Might Adventures, LP.

10 LOV 21 20 19 18

Contents

Chapter 1 Numbers 0 to 5
Mathematical Practice 6/Inquiry ... 1
Lesson 1 Count 1, 2, and 3 ... 2
Lesson 2 Read and Write 1, 2, and 3 ... 3
Lesson 3 Count 4 and 5 ... 4
Lesson 4 Read and Write 4 and 5 ... 5
Lesson 5 Read and Write Zero ... 6
Lesson 6 Equal To ... 7
Lesson 7 Greater Than ... 8
Lesson 8 Less Than ... 9
Lesson 9 Compare Numbers 0 to 5 ... 10
Lesson 10 One More ... 11
Lesson 11 Problem Solving Strategy: Draw a Diagram ... 12

Chapter 2 Numbers to 10
Mathematical Practice 1/Inquiry ... 13
Lesson 1 Numbers 6 and 7 ... 14
Lesson 2 Number 8 ... 15
Lesson 3 Read and Write 6, 7, and 8 ... 16
Lesson 4 Number 9 ... 17
Lesson 5 Number 10 ... 18
Lesson 6 Read and Write 9 and 10 ... 19
Lesson 7 Problem Solving Strategy: Act It Out ... 20
Lesson 8 Compare Numbers 0 to 10 ... 21
Lesson 9 One More With Numbers to 10 ... 22
Lesson 10 Ordinal Numbers to Fifth ... 23
Lesson 11 Ordinal Numbers to Tenth ... 24

Chapter 3 Numbers Beyond 10
Mathematical Practice 1/Inquiry ... 25
Lesson 1 Numbers 11 and 12 ... 26
Lesson 2 Numbers 13 and 14 ... 27
Lesson 3 Number 15 ... 28
Lesson 4 Numbers 16 and 17 ... 29
Lesson 5 Numbers 18 and 19 ... 30
Lesson 6 Number 20 ... 31
Lesson 7 Problem Solving Strategy: Draw a Diagram ... 32
Lesson 8 Count to 50 by Ones ... 33
Lesson 9 Count to 100 by Ones ... 34
Lesson 10 Count to 100 by Tens ... 35

Chapter 4 Compose and Decompose Numbers to 10
Mathematical Practice 2/Inquiry ... 36
Lesson 1 Make 4 and 5 ... 37
Lesson 2 Take Apart 4 and 5 ... 38
Lesson 3 Make 6 and 7 ... 39
Lesson 4 Take Apart 6 and 7 ... 40
Lesson 5 Problem Solving Strategy: Act It Out ... 41
Lesson 6 Make 8 and 9 ... 42
Lesson 7 Take Apart 8 and 9 ... 43
Lesson 8 Make 10 ... 44
Lesson 9 Take Apart 10 ... 45

Chapter 5 Addition
Mathematical Practice 6/Inquiry 46
Lesson 1 Addition Stories 47
Lesson 2 Use Objects to Add 48
Lesson 3 Use the + Symbol 49
Lesson 4 Use the = Symbol 50
Lesson 5 How Many in All? 51
Lesson 6 Problem Solving Strategy:
Write a Number Sentence 52
Lesson 7 Add to Make 10 53

Chapter 6 Subtraction
Mathematical Practice 6/Inquiry 54
Lesson 1 Subtraction Stories 55
Lesson 2 Use Objects to Subtract 56
Lesson 3 Use the − Symbol 57
Lesson 4 Use the = Symbol 58
Lesson 5 How Many Are Left? 56
Lesson 6 Problem Solving Strategy:
Write a Number Sentence 60
Lesson 7 Subtract to Take Apart 10 61

Chapter 7 Compose and Decompose Numbers 11 to 19
Mathematical Practice 2/Inquiry 62
Lesson 1 Make Numbers 11 to 15 63
Lesson 2 Take Apart Numbers 11 to 15 64
Lesson 3 Problem Solving Strategy:
Make a Table 65
Lesson 4 Make Numbers 16 and 19 66
Lesson 5 Take Apart Numbers 16 to 19 67

Chapter 8 Measurement
Mathematical Practice 5/Inquiry 68
Lesson 1 Compare Length 69
Lesson 2 Compare Height 70
Lesson 3 Problem Solving Strategy:
Guess, Check, and Revise 71
Lesson 4 Compare Weight 72
Lesson 5 Describe Length, Height, and Weight 73
Lesson 6 Compare Capacity 74

Chapter 9 Classify Objects
Mathematical Practice 7/Inquiry 75
Lesson 1 Alike and Different 76
Lesson 2 Problem Solving Strategy:
Use Logical Reasoning 77
Lesson 3 Sort by Size 78
Lesson 4 Sort by Shape 79
Lesson 5 Sort by Count 80

Chapter 10 Position
Mathematical Practice 1/Inquiry 81
Lesson 1 Above and Below 82
Lesson 2 In Front of and Behind 83
Lesson 3 Next to and Beside 84
Lesson 4 Problem Solving Strategy:
Act It Out 85

Chapter 11 Two-Dimensional Shapes

Mathematical Practice 7/Inquiry	86
Lesson 1 Squares and Rectangles	87
Lesson 2 Circles and Triangles	88
Lesson 3 Squares, Rectangles, Triangles, and Circles	89
Lesson 4 Hexagons	90
Lesson 5 Shapes and Patterns	91
Lesson 6 Shapes and Position	92
Lesson 7 Compose New Shapes	93
Lesson 8 Problem Solving Strategy: Use Logical Reasoning	94
Lesson 9 Model Shapes in the World	95

Chapter 12 Three-Dimensional Shapes

Mathematical Practice 7/Inquiry	96
Lesson 1 Spheres and Cubes	97
Lesson 2 Cylinders and Cones	98
Lesson 3 Compare Solid Shapes	99
Lesson 4 Problem Solving Strategy Act It Out	100
Lesson 5 Model Solid Shapes in Our World	101
Visual Kinesthetic Vocabulary©	VKV1

NAME _____ DATE _____

Chapter 1 Numbers 0 to 5
Inquiry of the Essential Question:

How do we show how many?

I see ...

I think ...

I know ...

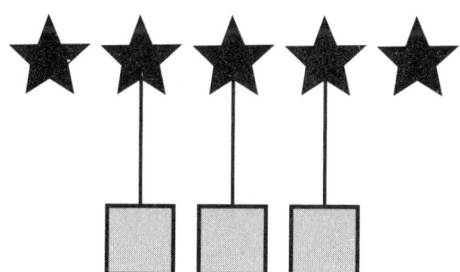

I see ...

I think ...

I know ...

I see ...

I think ...

I know ...

Questions I have...

Teacher Directions: Read the Essential Question for students. Have students echo read. Direct students to describe their observations, inferences, and prior knowledge of each math example. Encourage students to draw additional questions they may have. Scribe questions for students. Then have students share their ideas/questions with a peer.

Grade K • Chapter 1 *Numbers 0 to 5* 1

NAME _____ DATE _____

Lesson 1 Word Web

Count 1, 2, and 3

Trace the math word. Draw a picture story in each rectangle that shows the meaning of *count*.

```
 _ _ _ _ _ _ _ _ _ _
     count
 . . . . . . . . . .
```

Teacher Directions: Provide a description, explanation, or example of the new term using images or real objects. Have students say the letters aloud as they trace the math term. Direct students to draw two picture stories that represent their math term. Then encourage students to describe their pictures to a peer.

2 Grade K • Chapter 1 *Numbers 0 to 5*

NAME _____ DATE _____

Lesson 2 Word Identification

Read and Write 1, 2, and 3

Trace each word. Then match each number to the picture and word.

1

2      *one* (traced)

3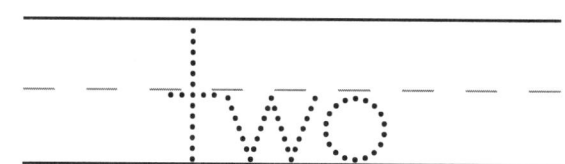

Draw 3 objects. Count and say how many.

Teacher Directions: Review counting 1, 2, and 3 using images or real objects. Have students say each number and then draw a line to match each number to a picture and word. Direct students to draw 3 objects, count, and then say how many. Then encourage students to describe their picture to a peer.

Grade K • Chapter 1 *Numbers 0 to 5*

NAME _____ DATE _____

Lesson 3 Number Identification

Count 4 and 5

Circle the pictures that show 5.

Draw a picture that shows 4.

Teacher Directions: Review counting to 4 and 5 using images or real objects. Have students identify and circle each example of 5. Direct students to draw a picture representing the number 4. Then encourage students to describe their picture to a peer.

4 Grade K • Chapter 1 Numbers 0 to 5

NAME _____   DATE _____

Lesson 4 Word Web
Read and Write 4 and 5

Trace the words. In the web, circle the correct word and number.

four 4 five 5	four 4 five 5

four

five

four 4 five 5	four 4 five 5

Teacher Directions: Provide a description, explanation, or example of the new terms using images or real objects. Have students say the letters aloud as they trace each math term. Direct students to circle the word and number that is represented by the picture. Then have students describe each picture to a peer. Encourage the use of a sentence frame such as: **There are [four/five] flowers.**

Grade K • Chapter 1 Numbers 0 to 5   5

NAME _____ DATE _____

Lesson 5 Word Identification

Read and Write Zero

Count the letters in the math word *zero*. Then say each letter.

| z-e-r-o |

Circle the math word *zero*.

a	e	b	o	z
g	j	u	t	e
q	e	y	r	r
z	e	r	o	o
x	i	g	p	z

Trace the number. Draw a picture to show what *zero* means.

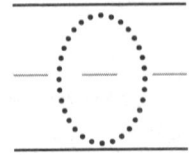

Teacher Directions: Provide a description, explanation, or example of the new term using images or real objects. Have students count the letters and then say each letter in the term. Have students identify, and circle two incidences of the math term. Direct students to trace the number. Then have them draw a picture to show what the term means. Finally encourage students to describe their picture to a peer.

6 Grade K • Chapter 1 *Numbers 0 to 5*

NAME _____ DATE _____

Lesson 6 Word Journal

Equal To

Trace the math word *equal*. Say the letters as you trace them.

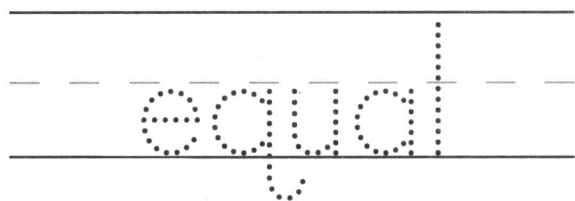

Draw a picture to show what *equal* means.

Teacher Directions: Provide a description, explanation, or example of the new term using images or real objects. Have students say the letters aloud as they trace the math term. Direct students to draw a picture representing their math term. Then encourage students to describe their picture to a peer.

NAME _____ DATE _____

Lesson 7 Concept Web

Greater Than

Use the web to show *greater than*.

glove    baseball

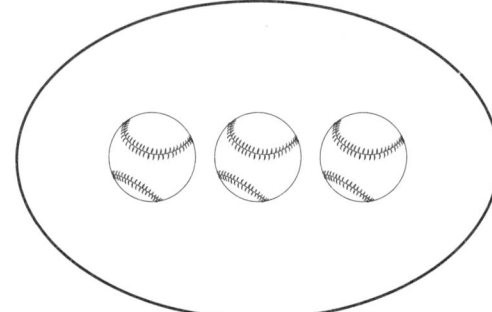

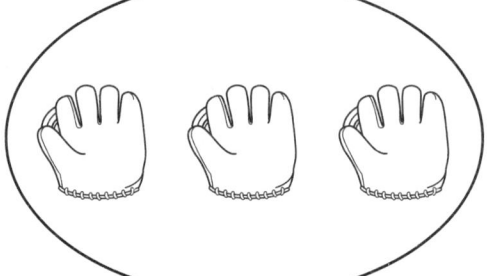

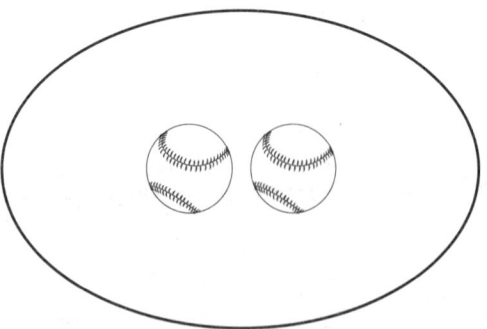   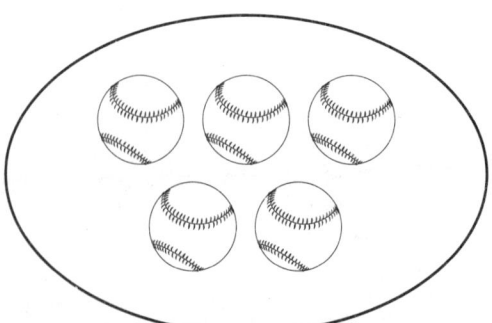

Teacher Directions: Provide a description, explanation, or example of the new term using images or real objects. Teach *glove* and *baseball* in the same manner. Have students count the number of gloves in the center oval. Then have them count the number of baseballs in one of the ovals. If the number of baseballs is greater than the number of gloves, have students draw a line from the baseballs to the gloves. Repeat for remaining sets of baseballs. Model and practice the sentence frame: **[Four/five] is greater than three.**

8 Grade K • Chapter I *Numbers 0 to 5*

Lesson 8 Concept Web

Less Than

Use the web to show *less than*.

 bee  flower

 **Teacher Directions:** Provide a description, explanation, or example of the new term using images or real objects. Teach *bee* and *flower* in the same manner. Have students count the number of bees in the center oval. Then have them count the number of flowers in one of the ovals. If the number of flowers is less than the number of bees, have students draw a line from the flowers to the bees. Repeat for remaining sets of flowers. Model and practice the sentence frame: **[Two/three] is less than four.**

Grade K • Chapter 1 *Numbers 0 to 5* **9**

NAME _____ DATE _____

Lesson 9 Sentence Frames
Compare Numbers 0 to 5

Count the objects in each sentence. Write the numbers. Read the sentences.

1.

_____ is less than _____

2.

_____ is greater than _____

3.

_____ is equal to _____

Teacher Directions: Provide a description, explanation, or example of the each term using images or real objects. Have students count the objects in the sentence frames and write the corresponding number below each group. Read each sentence frame and have students echo read. For example, **Two is less than three.** Then encourage students to read each sentence to a peer.

10 Grade K • Chapter 1 Numbers 0 to 5

NAME _____ DATE _____

Lesson 10 Word Web
One More

Trace the words. Draw a picture in each rectangle that shows the meaning of *one more*.

one more

Teacher Directions: Provide a description, explanation, or example of the term using images or real objects. Have students say the letters aloud as they trace the term. Direct students to draw two pictures that represent the words. Then encourage students to describe their pictures to a peer.

Grade K • Chapter 1 *Numbers 0 to 5* **11**

NAME _____ DATE _____

Lesson 11 Problem Solving
STRATEGY: Draw a Diagram

How many dogs have a bone?

Draw a Diagram

_____

- - - - -

_____

_____ **dogs** have **a bone**.

Teacher Directions: Provide a description, explanation, or example of the boldface terms using images or real objects. Review the problem solving strategy using the lesson example. Direct students to draw a line from each bone to a dog. Have students draw an X for each dog that has a bone and write the number. Have students write their answer in the restated question and read the answer sentence aloud.

12 Grade K • Chapter 1 *Numbers 0 to 5*

Chapter 2 Numbers to 10
Inquiry of the Essential Question:

What do numbers tell me?

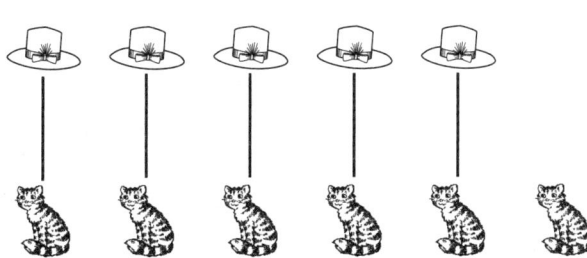

I see …

I think …

I know …

I see …

I think …

I know …

I see …

I think …

I know …

Questions I have…

Teacher Directions: Read the Essential Question for students. Have students echo read. Direct students to describe their observations, inferences, and prior knowledge of each math example. Encourage students to draw additional questions they may have. Scribe questions for students. Then have students share their ideas/questions with a peer.

Grade K • Chapter 2 *Numbers to 10*

NAME _____ DATE _____

Lesson 1 Word Web
Numbers 6 and 7

Trace the math words. Draw a picture story in each rectangle that shows the meaning of *six* and *seven*.

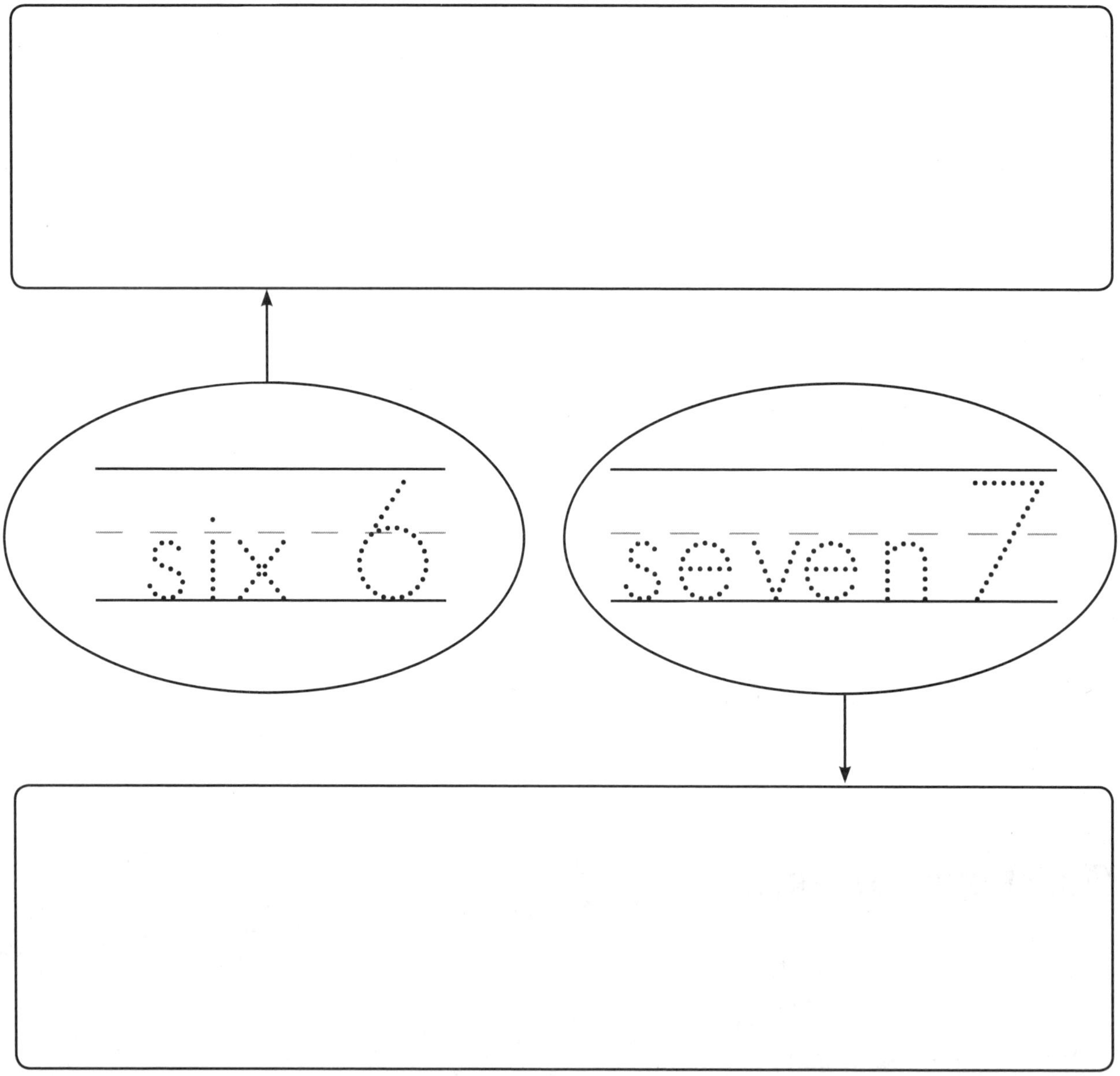

Teacher Directions: Provide a description, explanation, or example of the new terms using images or real objects. Have students say the letters aloud as they trace each math term. Direct students to draw a picture story that represents each math term. Then encourage students to describe their pictures to a peer.

NAME _____ DATE _____

Lesson 2 Vocabulary Word Identification
Number 8

Circle the math word *eight*.

eight	six	two	the
seven	and	eight	three
one	eight	count	eight
four	eight	five	or

Draw a picture of your math word.

Teacher Directions: Provide a description, explanation, or example of the new term using images or real objects. Have students identify and circle each incidence of the math term. Direct students to draw a picture representing their math term. Then encourage students to describe their picture to a peer.

Grade K • Chapter 2 *Numbers to 10* 15

Lesson 3 Number Identification
Read and Write 6, 7, and 8

Match each number to the picture and word.

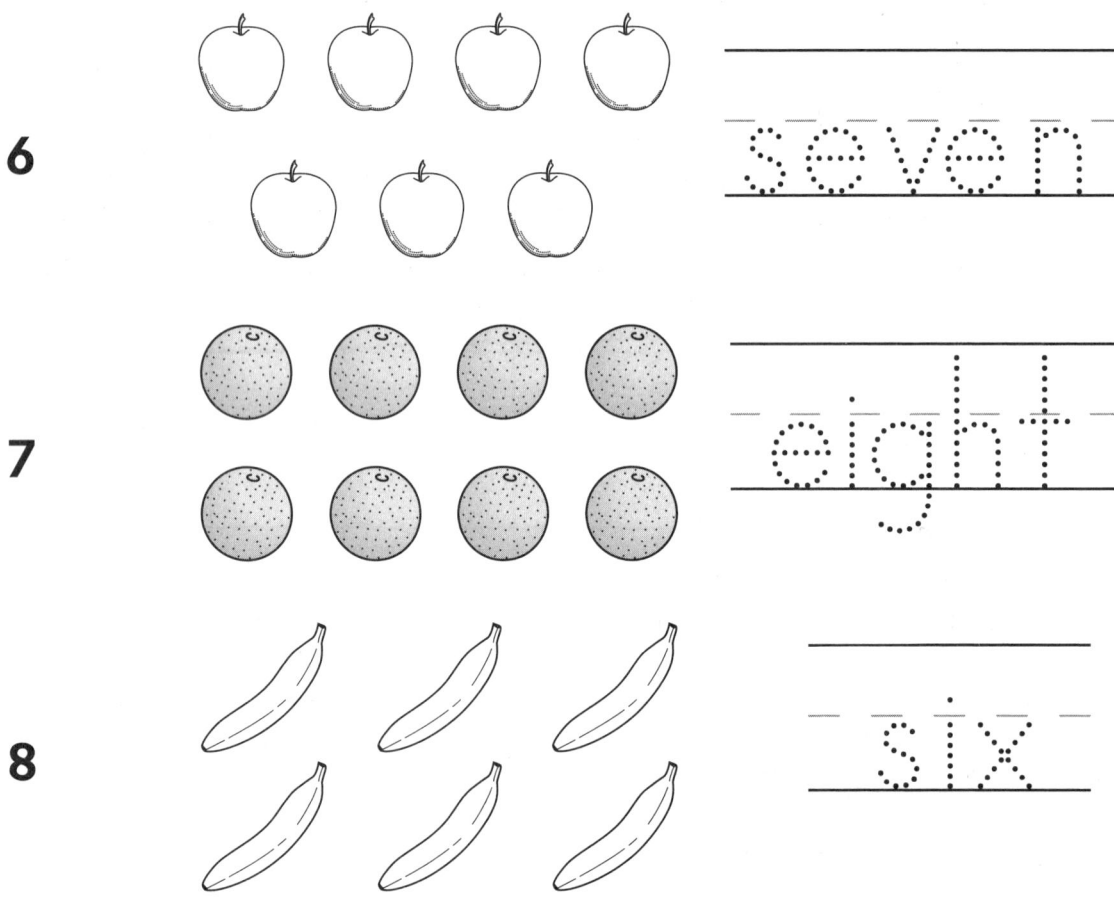

Draw 7 objects. Count and say how many.

Teacher Directions: Review counting to 6, 7, and 8 using images or real objects. Have students say each number then draw a line to match each number to a picture and word. Direct students to draw 7 objects. Then have them count and then say how many. Encourage students to describe their picture to a peer.

NAME _____ DATE _____

Lesson 4 Word Identification
Number 9

Count the letters in the math word *nine*. Then say each letter.

		n-i-n-e		

Circle the math word *nine*.

n	a	t	w	b
i	e	n	c	d
n	n	i	n	e
e	m	r	w	q
s	o	f	x	z

Trace the number. Draw a picture to show what *nine* means.

Teacher Directions: Provide a description, explanation, or example of the new term using images or real objects. Have students count the letters and then say each letter in the term. Have students identify and circle two incidences of the math term. Direct students to trace the number. Then have them draw a picture to show what the term means. Encourage students to describe their picture to a peer.

Grade K • Chapter 2 *Numbers to 10* 17

NAME _____ DATE _____

Lesson 5 Vocabulary Word Identification
Number 10

Circle the math word *ten*.

ten	two	to	the
ten	not	on	ten
no	if	three	nine
will	ten	ten	six

Draw a picture of your math word.

Teacher Directions: Provide a description, explanation, or example of the new term using images or real objects. Have students identify and circle each incidence of the math term. Direct students to draw a picture representing their math term. Then encourage students to describe their picture to a peer.

18 Grade K • Chapter 2 *Numbers to 10*

NAME _____ DATE _____

Lesson 6 Word Web
Read and Write 9 and 10

Trace the words. In the web, circle the correct word and number.

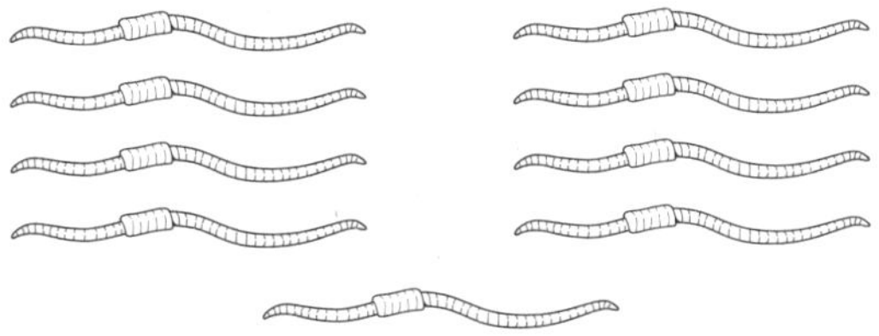

| nine 9   ten 10 | nine 9   ten 10 |

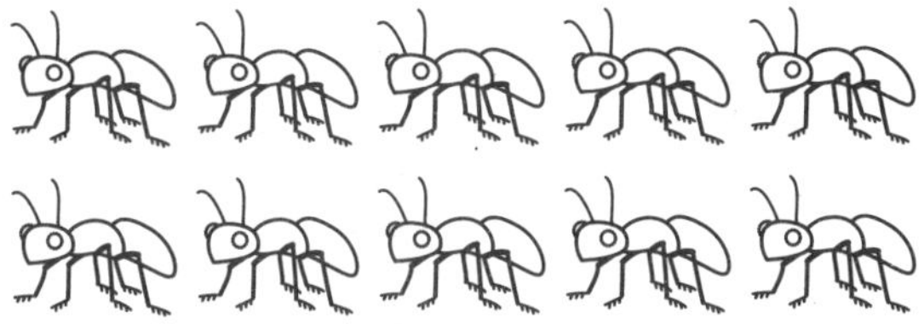

nine

ten

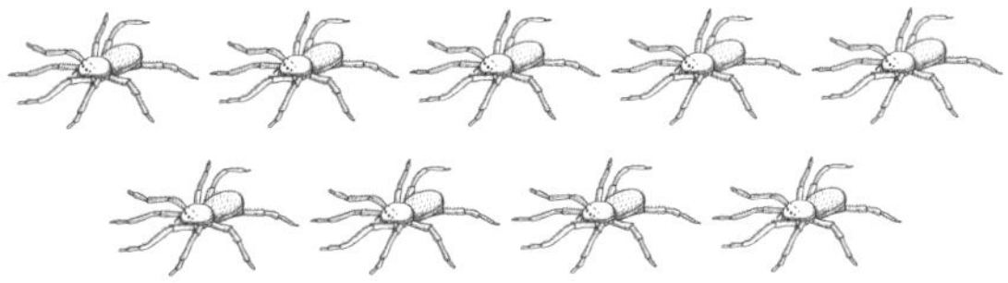

| nine 9   ten 10 | nine 9   ten 10 |

Teacher Directions: Provide a description, explanation, or example of the new terms using images or real objects. Have students say the letters aloud as they trace each math term. Direct students to circle the word and number that is represented by the picture. Then have students describe each picture to a peer. Encourage the use of a sentence frame such as: **There are nine worms.**

Grade K • Chapter 2 *Numbers to 10*

NAME _____ DATE _____

Lesson 7 Problem Solving
STRATEGY: Act It Out

How many animals are at the pond?

Act It Out

F	D	B
___	___	___
---	---	---
___	___	___

_____ **animals** are **at the pond.**

Teacher Directions: Point to each creature in the pond and introduce its name. Model and have students repeat. Provide a description, explanation, or example of the boldface terms in the question using images or real objects. Review the problem solving strategy using the lesson example. Direct students to act out the scene by having them mark each creature in the picture. For example, have them draw a *B* on each bee, a *D* on each dragonfly, and an *F* on each frog. Then have them count all the letters they wrote. Have students write their answer in the restated question and read the answer sentence aloud.

20 Grade K • Chapter 2 *Numbers to 10*

NAME _____ DATE _____

Lesson 8 Vocabulary Identification
Compare Numbers 0 to 10

Trace each term. Then draw a picture to show *less than*, *greater than*, and *equal to* 10.

less than	
greater than	
equal to	

Teacher Directions: Provide a description, explanation, or example of the terms using images or real objects. Direct students to draw a picture showing *less than*, *greater than*, or *equal to* 10. Then encourage students to describe their pictures to a peer. Suggest they use a sentence frame such as: **Four is less than ten.**

NAME _____ DATE _____

Lesson 9 Concept Web

One More with Numbers to 10

Trace the words. In the web, draw a picture to show a number and one more.

```
┌─────────────────┐        ┌─────────────────┐
│                 │        │                 │
│                 │        │                 │
│                 │        │                 │
│                 │        │                 │
│ 7 and one more  │        │ 4 and one more  │
└─────────────────┘        └─────────────────┘
          \                    /
           ┌──────────────────────┐
           │                      │
           │     one more         │
           │   (trace letters)    │
           │                      │
           └──────────────────────┘
          /                    \
┌─────────────────┐        ┌─────────────────┐
│                 │        │                 │
│                 │        │                 │
│                 │        │                 │
│                 │        │                 │
│ 3 and one more  │        │ 8 and one more  │
└─────────────────┘        └─────────────────┘
```

Teacher Directions: Provide a description, explanation, or example of the term using images or real objects. Have students say the letters aloud as they trace the math term. Direct students to draw a picture that represents the given number and one more. Then encourage students to describe each picture to a peer. Encourage the use of a sentence frame such as: **Seven and one more is eight.**

22 Grade K • Chapter 2 Numbers to 10

NAME _____   DATE _____

Lesson 10 Concept Web
Ordinal Numbers to Fifth

Trace the words. Trace the arrows to match each word to a car.

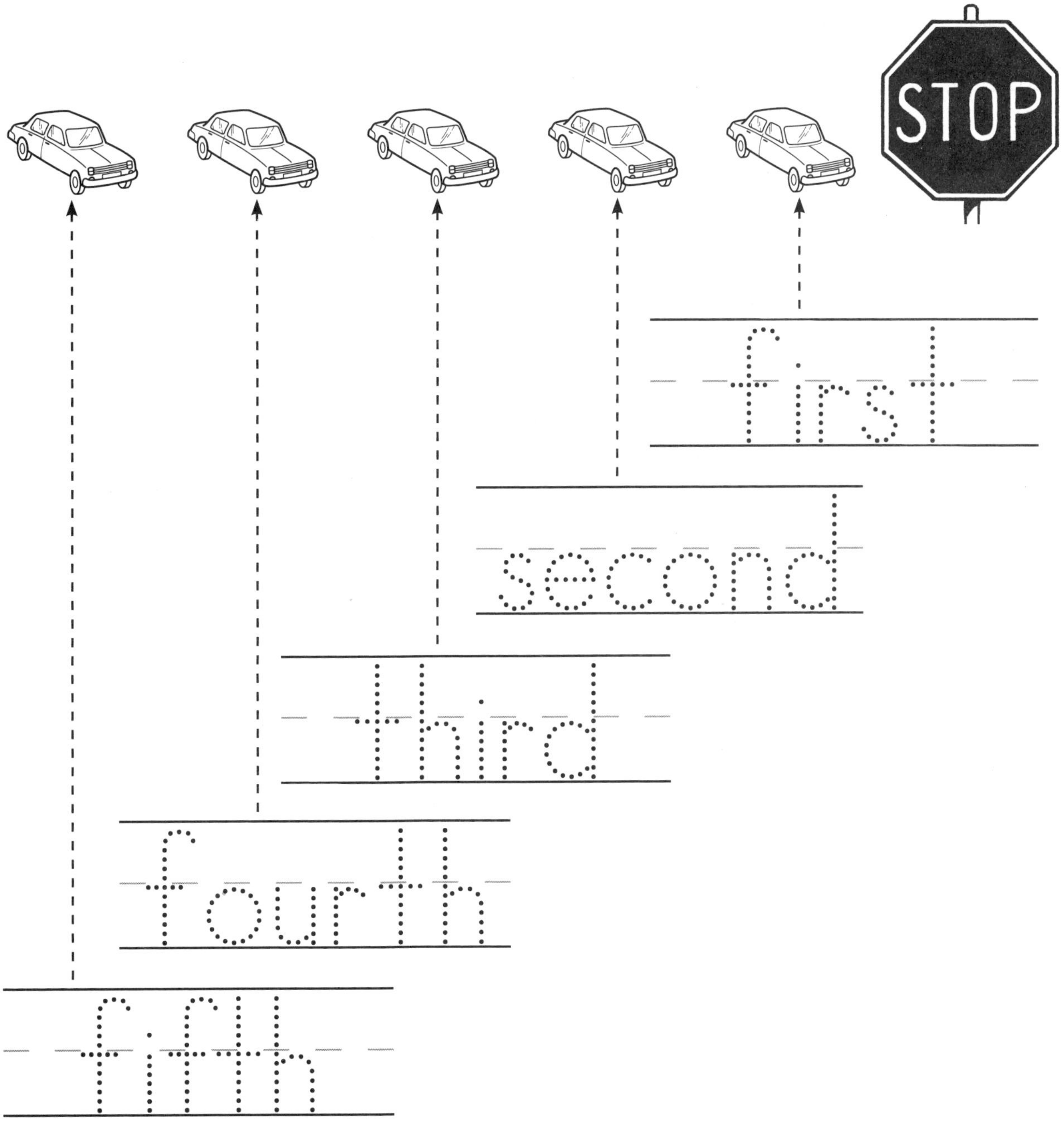

first

second

third

fourth

fifth

Teacher Directions: Provide a description, explanation, or example of the terms using images or real objects. Have students say the letters aloud as they trace each math term. Model how to pronounce each term and have students point to each term as you say it. Then have students repeat after you. Have students practice a complete sentence such as: **This car is fourth.**

Grade K • Chapter 2 Numbers to 10   **23**

NAME _____ DATE _____

Lesson 11 Concept Web

Ordinal Numbers to Tenth

Trace the math words. Draw a picture in each rectangle that shows the meaning of *ordinal numbers*.

```
┌─────────────────────────────────┐
│                                 │
│                                 │
└─────────────────────────────────┘
```

ordinal numbers

```
┌─────────────────────────────────┐
│                                 │
│                                 │
└─────────────────────────────────┘
```

Teacher Directions: Provide a description, explanation, or example of the new term using images or real objects. Have students say the letters aloud as they trace the math term. Direct students to draw two picture stories that represent their math term. Then encourage students to describe their pictures to a peer.

24 Grade K • Chapter 2 *Numbers to 10*

NAME _____ DATE _____

Chapter 3 Numbers Beyond 10
Inquiry of the Essential Question:

How can I show numbers beyond 10?

   20

I see ...

I think ...

I know ...

1	2	3	4	5	6	7	8	9	10
11	12	13	14	15	16	17	18	19	20
21	22	23	24	25	26	27	28	29	30
31	32	33	34	35	36	37	38	39	40
41	42	43	44	45	46	47	48	49	50
51	52	53	54	55	56	57	58	59	60
61	62	63	64	65	66	67	68	69	70
71	72	73	74	75	76	77	78	79	80
81	82	83	84	85	86	87	88	89	90
91	92	93	94	95	96	97	98	99	100

I see ...

I think ...

I know ...

🖐🖐 🖐🖐 🖐🖐 🖐🖐 🖐🖐
10   20   30   40   50

I see ...

I think ...

I know ...

Questions I have...

Teacher Directions: Read the Essential Question for students. Have students echo read. Direct students to describe their observations, inferences, and prior knowledge of each math example. Encourage students to draw additional questions they may have. Scribe questions for students. Then have students share their ideas/questions with a peer.

Grade K • Chapter 3 *Numbers Beyond 10* 25

NAME _____ DATE _____

Lesson 1 Word Web

Numbers 11 and 12

Trace the math words. Draw a picture in each rectangle that shows the meaning of *eleven* and *twelve*.

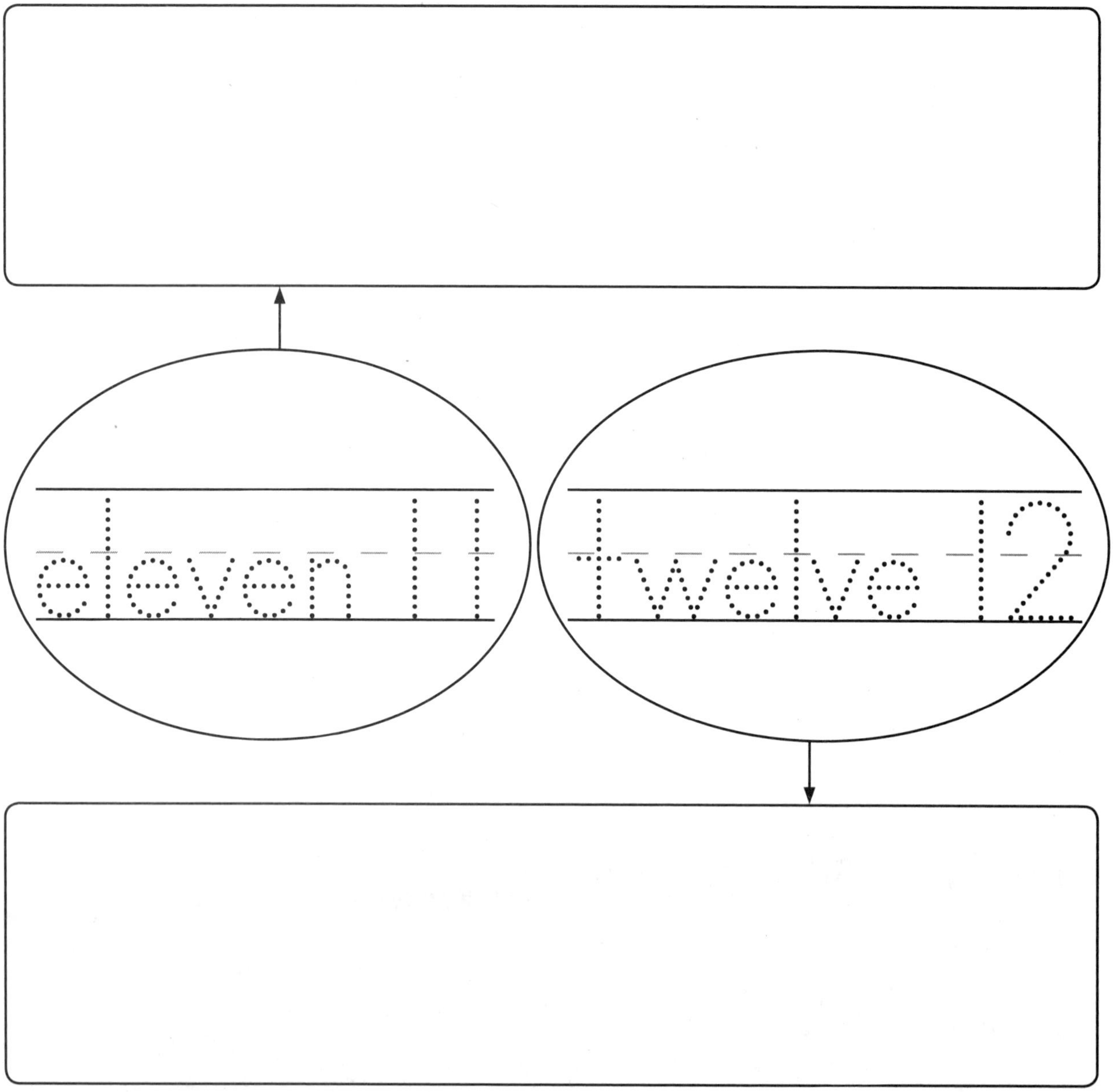

eleven 11   twelve 12

Teacher Directions: Provide a description, explanation, or example of the new terms using images or real objects. Have students say the letters aloud as they trace each math term. Direct students to draw a picture that represents each math term. Then encourage students to describe their pictures to a peer.

26 Grade K • Chapter 3 *Numbers Beyond 10*

NAME _____ DATE _____

Lesson 2 Number Identification
Numbers 13 and 14

Match each number to the picture.

12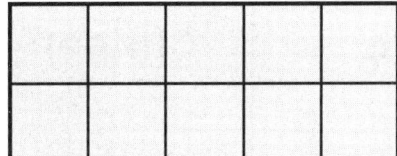

13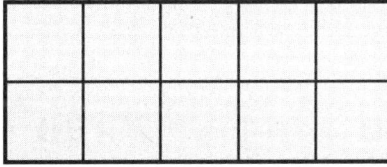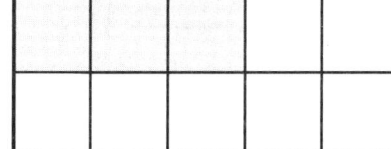

14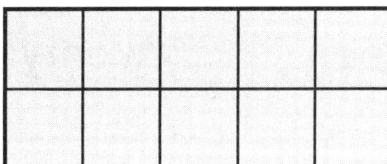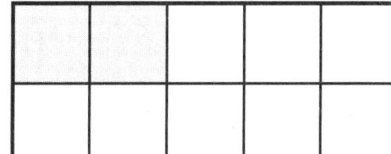

Draw 13 objects. Count and say how many.

 **Teacher Directions:** Review counting to 14 using images or real objects. Have students say each number then draw a line to match each number to a picture. Direct students to draw 13 objects. Then have them count and say how many. Encourage students to describe their picture to a peer.

Grade K • Chapter 3 *Numbers Beyond 10*

Lesson 3 Vocabulary Word Identification
Number 15

Circle the math word *fifteen*.

ten	five	fifteen	twelve
eleven	nine	fifteen	sixteen
thirteen	one	three	fifteen
fifteen	fourteen	twenty	seven

Draw a picture of your math word.

Teacher Directions: Provide a description, explanation, or example of the new term using images or real objects. Have students identify, and circle each incidence of the math term. Direct students to draw a picture representing their math term. Then encourage students to describe their picture to a peer.

NAME _____ DATE _____

Lesson 4 Word Identification
Numbers 16 and 17

Trace then match each number to the number word.

15

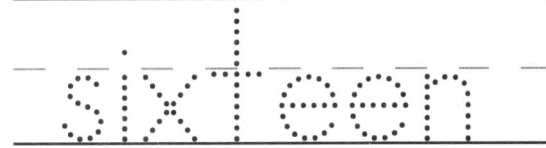

16

17

Draw 17 objects. Count and say how many.

Teacher Directions: Review counting to 17 using images or real objects. Have students say each number, trace and say the word, then draw a line to match each number to the number word. Direct students to draw 17 objects. Then have them count and say how many. Encourage students to describe their picture to a peer.

Grade K • Chapter 3 Numbers Beyond 10

NAME ______________________ DATE ____________

Lesson 5 Word Web

Numbers 18 and 19

Trace the math words.

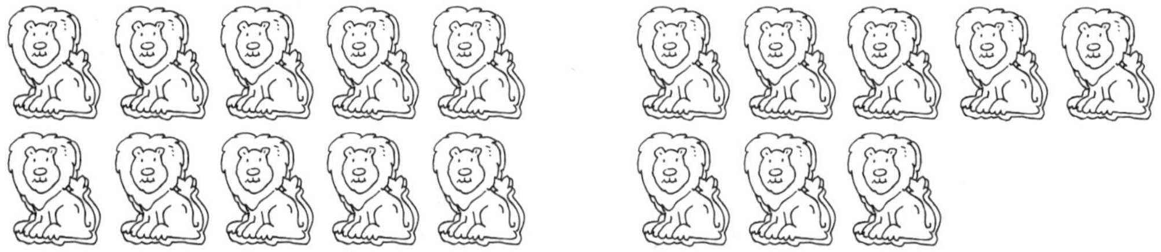

eighteen 18

nineteen 19

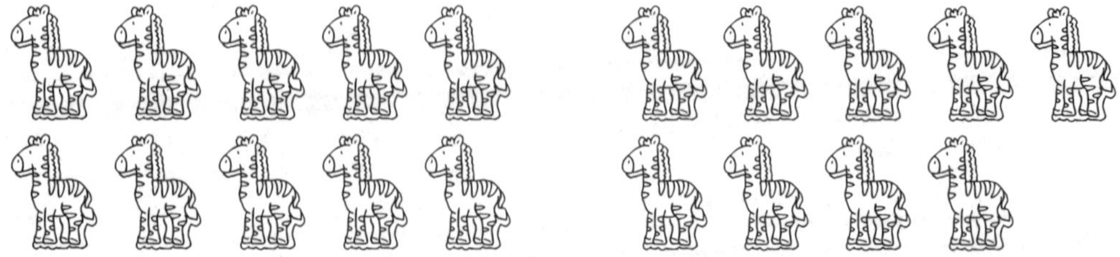

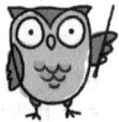

Teacher Directions: Provide a description, explanation, or example of the new terms using images or real objects. Have students say the letters aloud as they trace each math term. Direct students to circle groups of ten then count on to the corresponding number. Then encourage students to describe the pictures to a peer using a sentence frame such as: **There are 19 giraffes.**

NAME _____ DATE _____

Lesson 6 Word Recognition
Number 20

Count the letters in the math word *twenty* then say each letter.

		t-w-e-n-t-y			

Circle the math word *twenty*.

x	t	h	i	u	v
t	w	e	n	t	y
b	e	o	l	p	u
m	n	p	c	z	j
k	t	k	u	x	c
n	y	o	p	e	l

Trace the number. Draw a picture to show what *twenty* means.

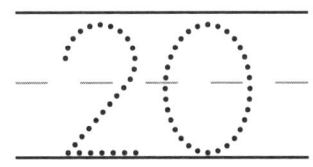

Teacher Directions: Provide a description, explanation, or example of the new term using images or real objects. Have students count the letters then say each letter in the term. Have students identify and circle two incidences of the math term. Direct students to trace the number, then draw a picture to show what the term means. Then encourage students to describe their picture to a peer.

Grade K • Chapter 3 *Numbers Beyond 10* **31**

NAME _____ DATE _____

Lesson 7 Problem Solving
STRATEGY: Draw a Diagram

How many **frogs** are **at the pond**?

Draw to Solve

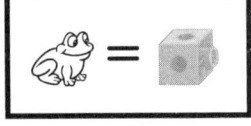

____ **frogs** are **at the pond**.

Teacher Directions: Provide a description, explanation, or example of the boldface terms using images or real objects. Review the problem solving strategy using the lesson example. Have students place a cube on each frog. Direct students to draw an X for each frog/cube, count the Xs, and then write the number. Have students write their answer in the restated question and read the answer sentence aloud.

32 Grade K • Chapter 3 *Numbers Beyond 10*

NAME _____ DATE _____

Lesson 8 Word Journal
Count to 50 by Ones

Trace the math word *count*. Say the letters as you trace them.

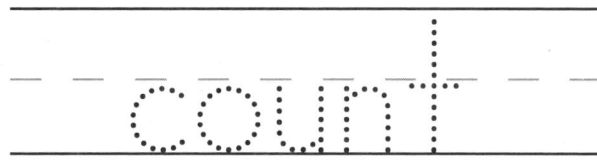

Draw a picture to show what *count* means.

Teacher Directions: Provide a description, explanation, or example of the term using images or real objects. Have students say the letters aloud as they trace the math term. Direct students to draw a picture representing their math term. Then encourage students to describe their picture to a peer.

NAME _____ DATE _____

Lesson 9 Concept Web

Count to 100 by Ones

Trace the word. Circle the shapes that show *ones*.

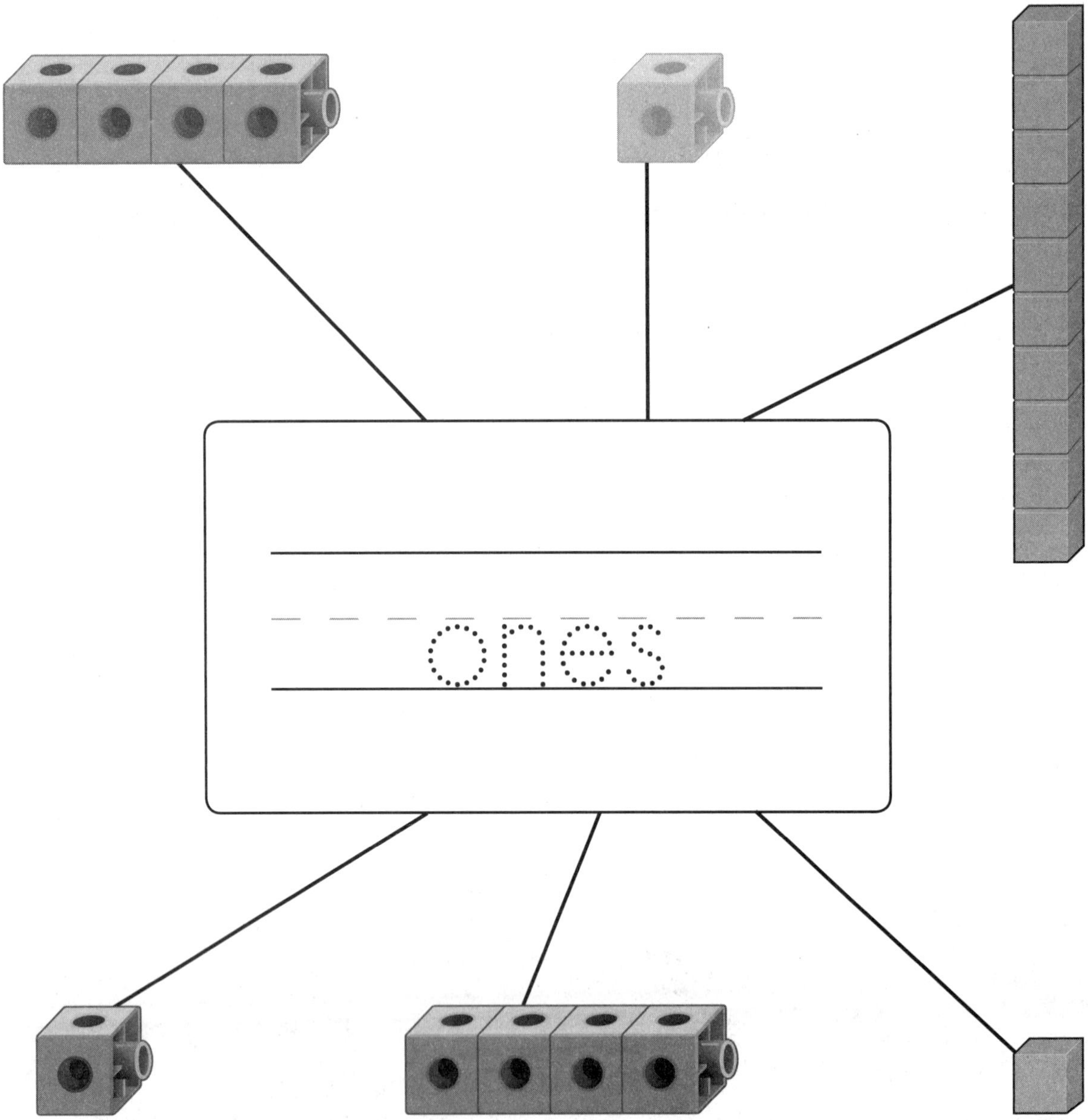

_____ ones _____

Teacher Directions: Provide a description, explanation, or example of the term using images or real objects. Have students say the letters aloud as they trace the math term. Have students circle each example of *ones*. Then encourage them to describe each picture to a peer.

34 Grade K • Chapter 3 *Numbers Beyond 10*

Lesson 10 Chart

Count to 100 by Tens

Trace the word. Circle all the *tens* on the chart.

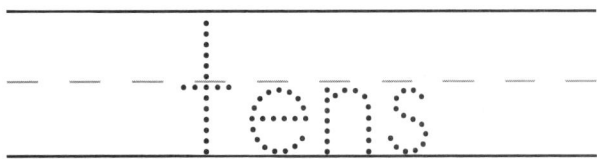

1	2	3	4	5	6	7	8	9	10
11	12	13	14	15	16	17	18	19	20
21	22	23	24	25	26	27	28	29	30
31	32	33	34	35	36	37	38	39	40
41	42	43	44	45	46	47	48	49	50
51	52	53	54	55	56	57	58	59	60
61	62	63	64	65	66	67	68	69	70
71	72	73	74	75	76	77	78	79	80
81	82	83	84	85	86	87	88	89	90
91	92	93	94	95	96	97	98	99	100

Teacher Directions: Provide a description, explanation, or example of the term using images or real objects. Have students say the letters aloud as they trace the math term. Have students circle each example of *tens*. Then encourage partners to use the chart to help them count to 100 by tens.

Grade K • Chapter 3 Numbers Beyond 10

NAME _____ DATE _____

Chapter 4 Compose and Decompose Numbers to 10

Inquiry of the Essential Question:

How can we show a number in other ways?

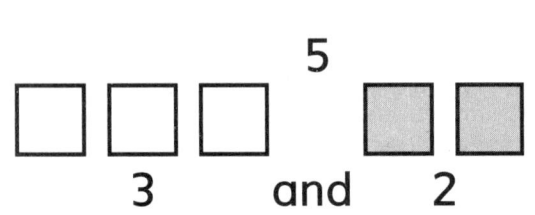

I see ...

I think ...

I know ...

I see ...

I think ...

I know ...

I see ...

I think ...

I know ...

Questions I have...

Teacher Directions: Read the Essential Question for students. Have students echo read. Direct students to describe their observations, inferences, and prior knowledge of each math example. Encourage students to draw additional questions they may have. Scribe questions for students. Then have students share their ideas/questions with a peer.

Lesson 1 Sum Identification
Make 4 and 5

Circle all the ways to make 4.

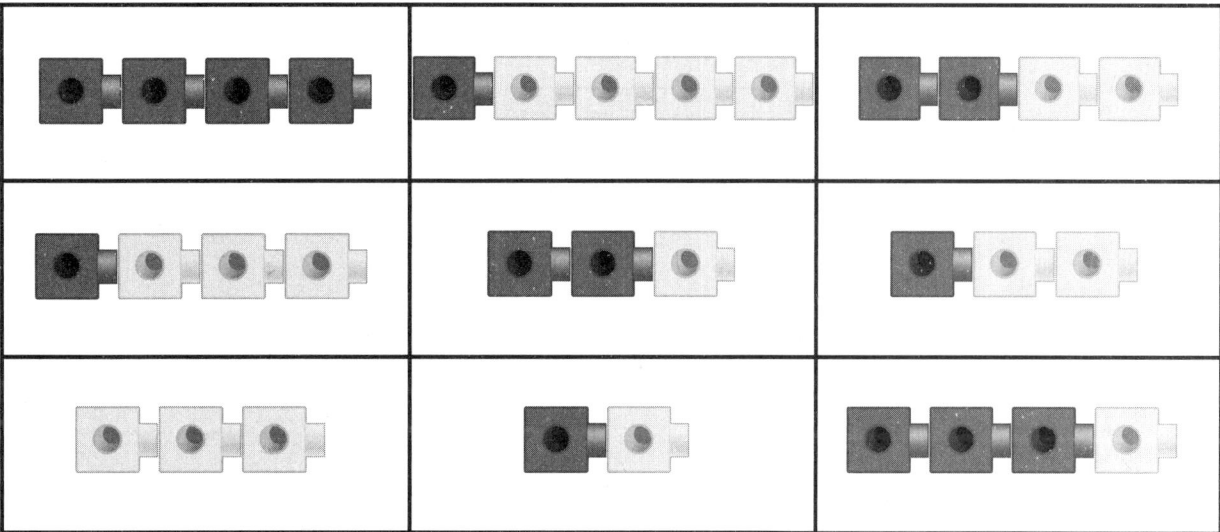

Draw a picture to show one way to make 5.

Teacher Directions: Use manipulatives to show various ways to make 4 and 5. Model an addition sentence for each and have students repeat it. Have students circle each example that shows a sum of 4. Then direct students to draw a picture representing a way to make 5. Encourage students to describe their picture to a peer.

NAME _____ DATE _____

Lesson 2 Word Journal

Take Apart 4 and 5

Trace the math word *take apart*. Say the letters as you trace them.

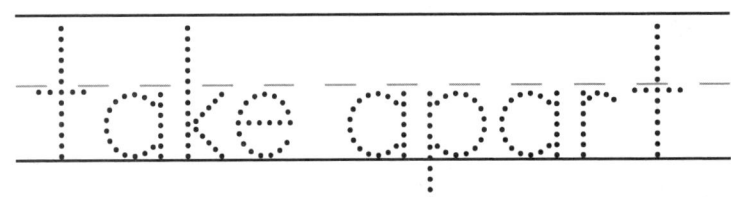

Draw a picture to show what *take apart* means.

Teacher Directions: Provide a description, explanation, or example of the new term using images or real objects. Have students say the letters aloud as they trace the math term. Direct students to draw a picture representing their math term. Then encourage students to describe their picture to a peer.

NAME _____ DATE _____

Lesson 3 Sum Identification
Make 6 and 7

Circle all the ways to make 7.

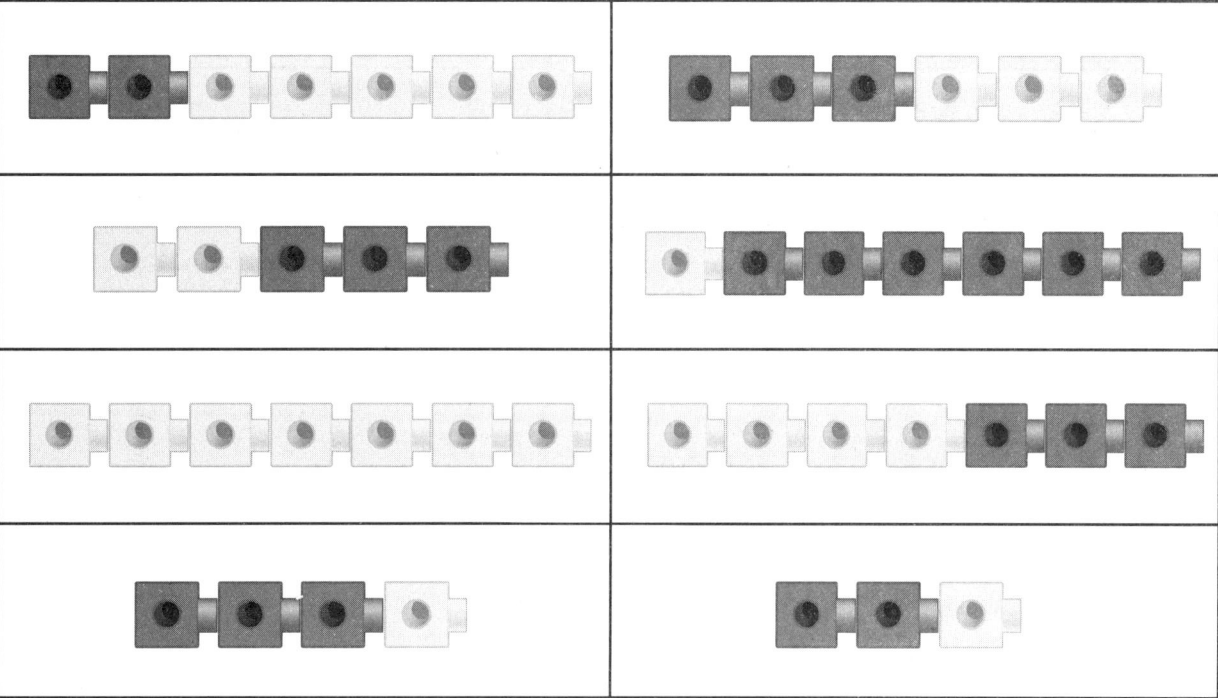

Draw a picture to show one way to make 6.

Teacher Directions: Use manipulatives to show various ways to make 6 and 7. Model an addition sentence for each and have students repeat it. Have students circle each example that shows a sum of 7. Then direct students to draw a picture representing a way to make 6. Encourage students to describe their picture to a peer.

Grade K • Chapter 4 Compose and Decompose Numbers to 10   **39**

NAME _____ DATE _____

Lesson 4 Word Web

Take Apart 6 and 7

Trace the math words. Draw a line to the correct examples for taking apart each number.

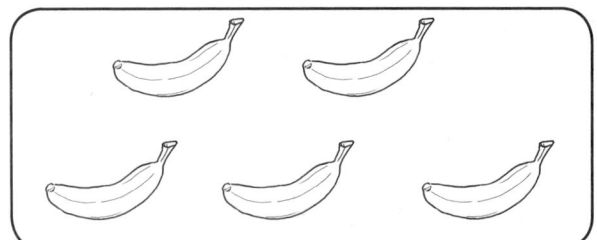

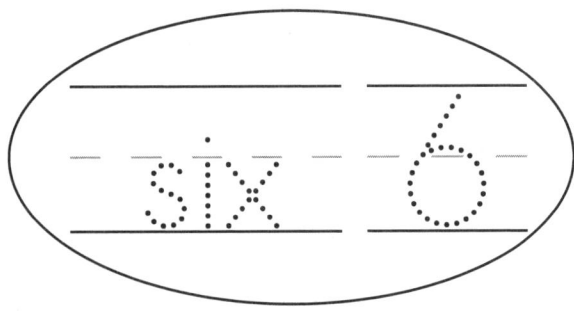
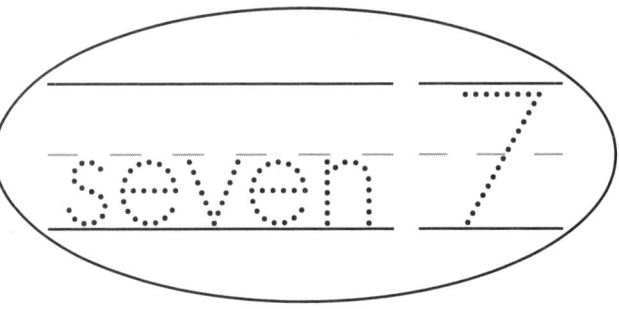

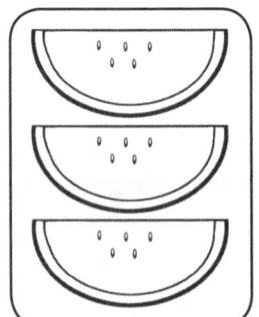
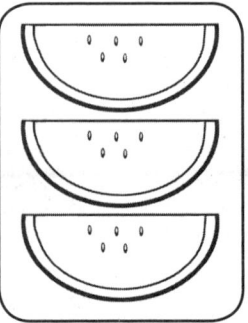

Teacher Directions: Provide a description, explanation, or example of the term *take apart* using images or real objects. Have students say the letters aloud as they trace each number and word. Direct students to draw a line from the word to the picture that represents how to take apart each number. Encourage students to tell a partner how each group was taken apart. For example: **Six is three and three.**

40 Grade K • Chapter 4 Compose and Decompose Numbers to 10

Lesson 5 Problem Solving
STRATEGY: Act It Out

How do you make the number 7?

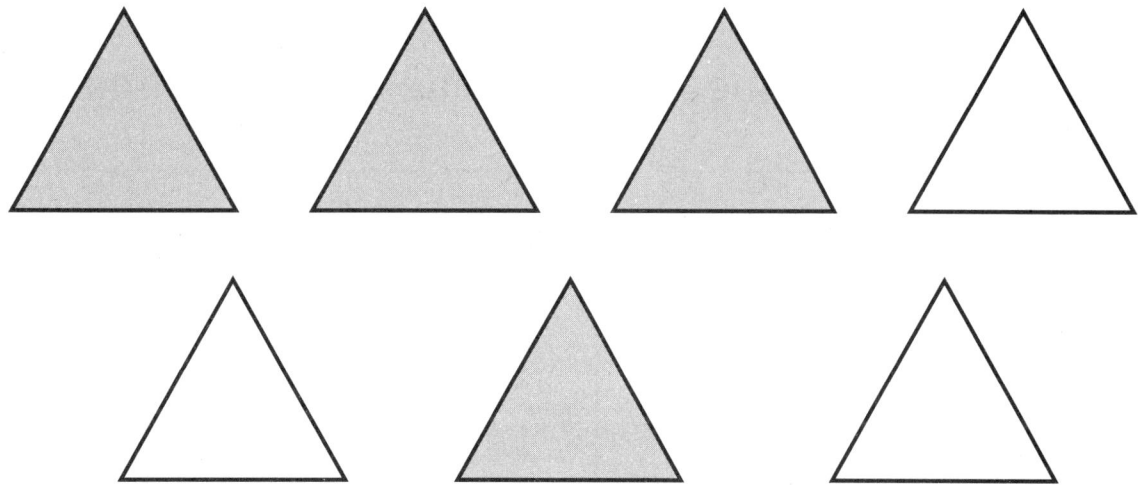

Act It Out

_____ and _____ make _____.

Teacher Directions: Provide a description, explanation, or example of the boldface terms using images or real objects. Review the problem solving strategy using the lesson example. Direct students to use manipulatives to act out the problem. Have students write their answer in the restated question and read the answer sentence aloud.

Grade K • Chapter 4 *Compose and Decompose Numbers to 10* **41**

NAME _____ DATE _____

Lesson 6 Vocabulary Word Identification
Make 8 and 9

Circle the math word *nine*.

time	vine	nine	dime
like	dine	nine	nine
fine	shine	mine	line
nine	spine	nine	twine

Draw a way to make 8.

Teacher Directions: Review counting up to 9 using images or real objects. Then use manipulatives to practice ways to make 8 and 9. Have students identify and circle each incidence of the word *nine*. Direct students to draw a picture representing a way to make 8. Then encourage students to describe their picture to a peer.

Lesson 7 Number Identification
Take Apart 8 and 9

Match each number to the picture.

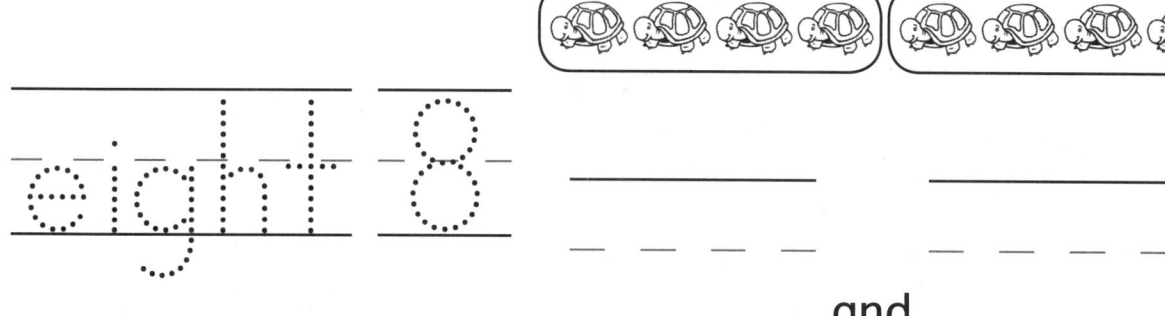

eight 8

_____ _____
_ _ _ _

_____ and _____

nine 9

_____ _____
_ _ _ _

_____ and _____

Draw 8 objects. Count and say how many.

Teacher Directions: Review counting up to 9 using images or real objects. Then use manipulatives to practice ways to make 8 and 9. Have students count the objects and write the numbers that make 8 and 9. Then have them draw a line to match each number to the corresponding groups of objects. Direct students to draw 8 objects, count, then say how many. Encourage students to describe their picture to a peer.

Grade K • Chapter 4 Compose and Decompose Numbers to 10   43

NAME _____ DATE _____

Lesson 8 Word Identification

Make 10

Count the letters in the math word *ten*. Then say each letter.

t-e-n

Circle the math word *ten*.

w	a	t	e	n
i	e	n	c	d
t	v	i	n	d
e	m	r	s	q
n	o	f	g	z

Trace the number. Draw a picture to show how to make ten.

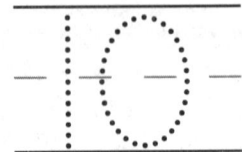

Teacher Directions: Provide a description, explanation, or example of the term using images or real objects. Have students count the letters and then say each letter in the term. Have students identify and circle two incidences of the math term. Direct students to trace the number. Then have them draw a picture to show how to make ten. Encourage students to describe their picture to a peer.

44 Grade K • Chapter 4 Compose and Decompose Numbers to 10

NAME _____ DATE _____

Lesson 9 Word Web

Take Apart 10

Trace the word. In the web, draw a line from the word to the ways to take apart ten.

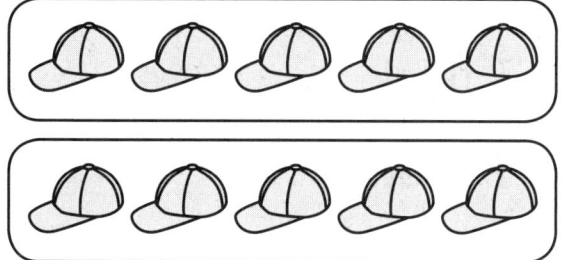
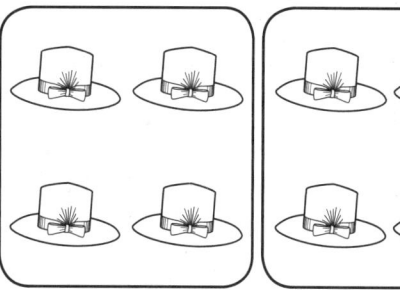

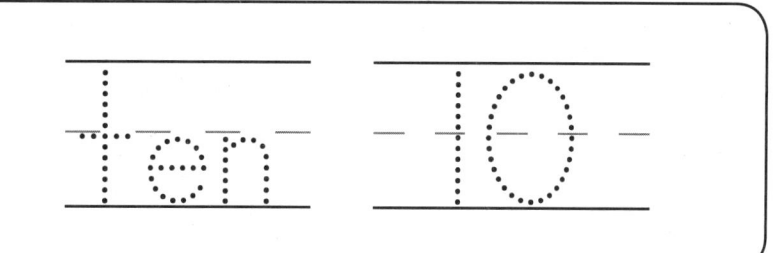

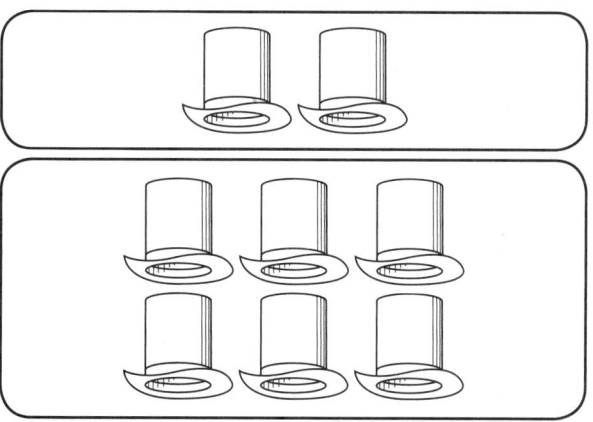
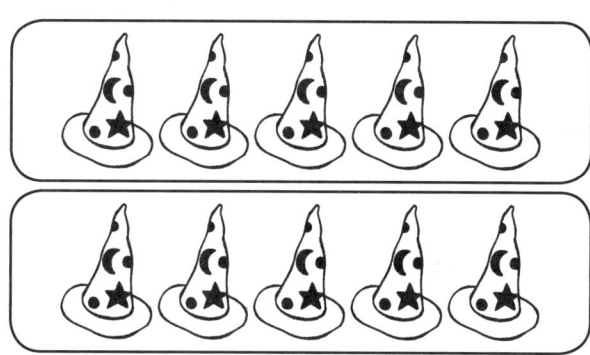

Teacher Directions: Provide a description, explanation, or example of the term using images or real objects. Have students say the letters aloud as they trace the math term. Direct students to draw a line to each picture that shows a way to take apart 10. Then encourage students to describe each picture to a peer.

Grade K • Chapter 4 *Compose and Decompose Numbers to 10* **45**

NAME _____ DATE _____

Chapter 5 Addition

Inquiry of the Essential Question:

How can I use objects to add?

1 + 9 = 10	9 + 1 = 10	I see ...
2 + 8 = 10	8 + 2 = 10	I think ...
3 + 7 = 10	7 + 3 = 10	
4 + 6 = 10	6 + 4 = 10	I know ...
5 + 5 = 10		

7 + 3 = 10
↑   ↑
plus equals

I see ...

I think ...

I know ...

I see ...

I think ...

I know ...

Questions I have...

Teacher Directions: Read the Essential Question for students. Have students echo read. Direct students to describe their observations, inferences, and prior knowledge of each math example. Encourage students to draw additional questions they may have. Scribe questions for students. Then have students share their ideas/questions with a peer.

46 Grade K • Chapter 5 *Addition*

Lesson 1 Word Journal
Addition Stories

Trace the math word *join*. Say the letters as you trace them.

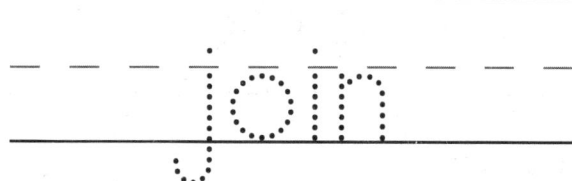

Draw a picture to show what *join* means.

Teacher Directions: Provide a description, explanation, or example of the new term using images or real objects. Have students say the letters aloud as they trace the math term. Direct students to draw a picture representing their math term. Then encourage students to describe their picture to a peer.

Lesson 2 Word Recognition
Use Objects to Add

Count the letters in the math word *add*. Then say each letter.

a-d-d

Circle the math word *add*.

b	y	v	o	p
a	u	y	e	w
d	k	a	d	d
d	q	y	r	r
i	b	m	s	a

Trace the word. Draw a picture to show what *add* means.

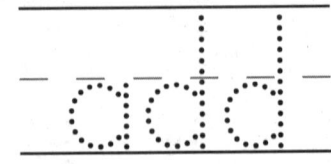

Teacher Directions: Provide a description, explanation, or example of the new term using images or real objects. Have students count the letters then say each letter in the term. Have students identify and circle two incidences of the math term. Direct students to trace the word and then draw a picture to show what the term means. Then encourage students to describe their picture to a peer.

48 Grade K • Chapter 5 Addition

NAME _____ DATE _____

Lesson 3 Symbol Identification
Use the + Symbol

Trace the plus signs. Match.

 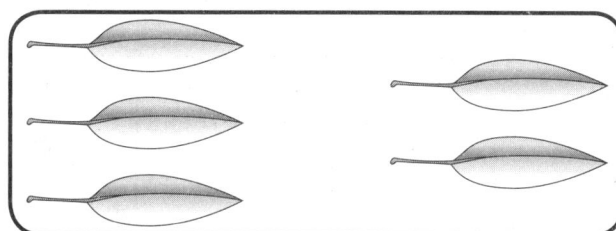

 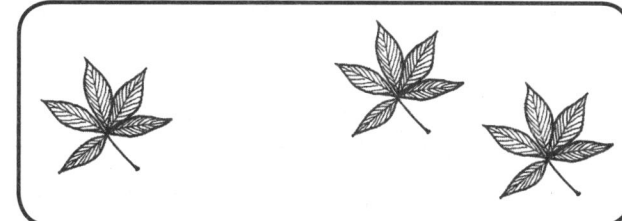

 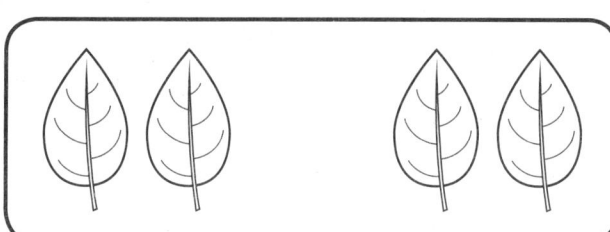

Draw 3 objects. Draw 1 more object. Write the numbers on each line. Trace the plus sign.

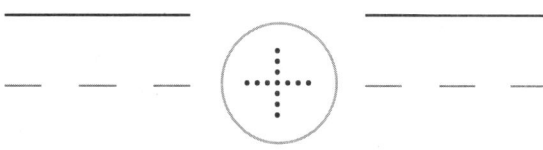

 **Teacher Directions:** Provide a description, explanation, or example of the new term using images or real objects. Have students say each expression on the left and draw a line to match each expression to a picture. Direct students to draw 3 objects and then 2 more objects. Then have them write the numbers on the lines and trace the plus sign. Encourage students to describe their picture to a peer.

Grade K • Chapter 5 Addition

NAME _____ DATE _____

Lesson 4 Vocabulary Word Identification

Use the = Symbol

Circle the math word *equals* in the boxes below.

add	join	equals	plus
ten	one	two	equals
join	four	equals	six
equals	eight	three	seven

Trace the symbol. Draw a picture to show what *equals* means.

— — ⋮⋮ — —

Teacher Directions: Provide a description, explanation, or example of the new term using images or real objects. Have students identify and circle each incidence of the math term. Direct students to trace the equals sign and then draw a picture representing their math term. Encourage students to describe their picture to a peer.

50 Grade K • Chapter 5 Addition

Lesson 5 Word Web

How Many in All?

Trace the math words. Draw a picture story in each rectangle that shows the meaning of *in all*.

in all

Teacher Directions: Provide a description, explanation, or example of the term using images or real objects. Have students say the letters aloud as they trace the math term. Direct students to draw two picture stories that represent their math term. Then encourage students to describe their pictures to a peer.

Grade K • Chapter 5 Addition 51

NAME _____ DATE _____

Lesson 6 Problem Solving
STRATEGY: Write a Number Sentence

How many are there **in all**?

Write a Number Sentence

___ + ___ = ___

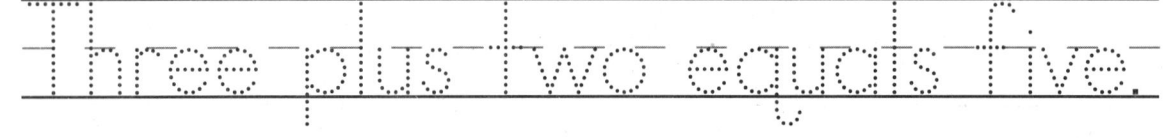

Three plus two equals five.

There are ____ in all.

Teacher Directions: Provide a description, explanation, or example of the boldface terms using images or real objects. Review the problem solving strategy using the lesson example. Have students count aloud as they draw an X on each apple inside the basket. Then have them write the number on the first line. Repeat for the apples outside the basket. Have students trace the plus sign and equals sign and then say the number sentence. Then have them trace the words in the sentence. Explain the difference between a number sentence and a sentence that uses words. Finally, have students write their answer in the restated question and read the answer sentence aloud.

52 Grade K • Chapter 5 Addition

NAME _____ DATE _____

Lesson 7 Concept Web
Add to Make 10

Trace the words and symbols.

- join
- plus +
- add
- equals =
- in all

Draw a picture of one of the words.

Teacher Directions: Review each term and symbol by providing a description, explanation, or example using images or real objects. Have students say the letters aloud as they trace each math term and symbol. Direct students to draw a picture that represent one of the math terms. Then encourage students to describe their picture to a peer.

Grade K • Chapter 5 *Addition* 53

NAME _____ DATE _____

Chapter 6 Subtraction

Inquiry of the Essential Question:

How can I use objects to subtract?

10 − 1 = 9   I see …
10 − 2 = 8
10 − 3 = 7   I think …
10 − 4 = 6
10 − 5 = 5   I know …
10 − 6 = 4
10 − 7 = 3
10 − 8 = 2
10 − 9 = 1
10 − 10 = 0

10 − 6 = 4

I see …

I think …

I know …

7 − 3 = 4

I see …

I think …

I know …

Questions I have…

 **Teacher Directions:** Read the Essential Question for students. Have students echo read. Direct students to describe their observations, inferences, and prior knowledge of each math example. Encourage students to draw additional questions they may have. Scribe questions for students. Then have students share their ideas/questions with a peer.

Lesson 1 Word Journal

Subtraction Stories

Trace the math term *take away*. Say the letters as you trace them.

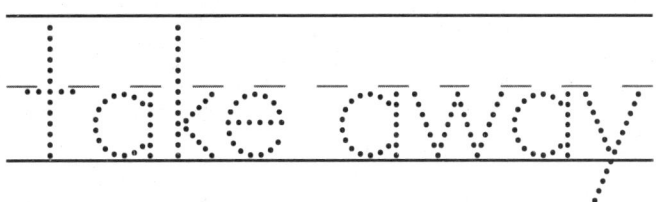

Draw a picture to show what *take away* means.

Teacher Directions: Provide a description, explanation, or example of the new term using images or real objects. Have students say the letters aloud as they trace the math term. Direct students to draw a picture representing their math term. Then encourage students to describe their picture to a peer.

Grade K • Chapter 6 *Subtraction*

NAME _____ DATE _____

Lesson 2 Word Recognition
Use Objects to Subtract

Circle the math word *subtract* in the boxes below.

add	join	five	plus
subtract	one	subtract	equals
ten	four	nine	subtract
subtract	eight	three	seven

Trace the word. Draw a picture to show what *subtract* means.

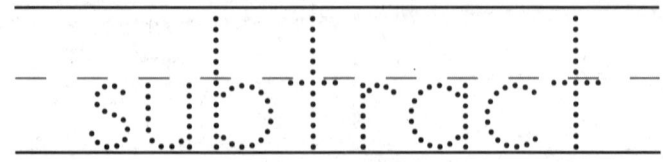

Teacher Directions: Provide a description, explanation, or example of the new term using images or real objects. Have students identify and circle each incidence of the math term. Direct students to draw a picture representing their math term. Then encourage students to describe their picture to a peer.

NAME _____ DATE _____

Lesson 3 Symbol Identification

Use the – Symbol

Trace each minus sign. Match.

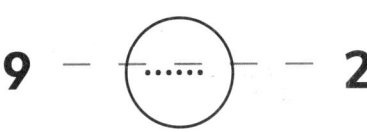   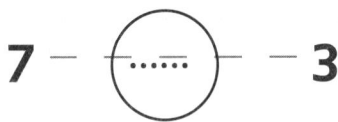

9 – ⊙ – 2   7 – ⊙ – 3

Draw 8 objects. Draw an X over 4 objects. Write the numbers on each line. Trace the minus sign.

____ ⊙ ____   ____ is ____.

Teacher Directions: Provide a description, explanation, or example of the new term using images or real objects. Have students trace each minus sign, say each expression, and draw a line to match each expression to a picture. Have students use phrases such as: **Nine minus two** as they complete their matches. Direct students to draw 8 objects and then cross out 4 objects. Then have them write the numbers on the lines and trace the minus sign. Have them count how many are left and write the number on the last line. Then encourage students to describe their pictures to a peer.

Grade K • Chapter 6 *Subtraction* 57

NAME _____ DATE _____

Lesson 4 Vocabulary Word Identification

Use the = Symbol

Count the letters in the math word *equals*.
Then say each letter.

e-q-u-a-l-s

Circle the math word *equals*.

f	l	u	r	p	e
w	n	o	v	k	q
q	w	e	r	t	u
u	i	n	n	o	a
h	i	f	c	v	l
e	q	u	a	l	s

Trace the word. Draw a picture to show what *equals* means.

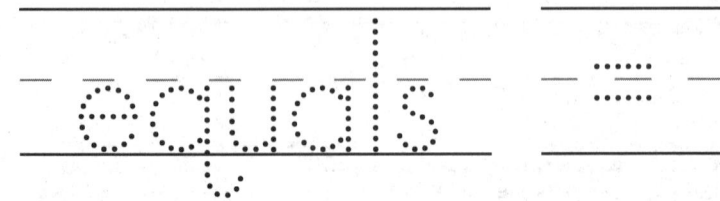

Teacher Directions: Provide a description, explanation, or example of the new term using images or real objects. Have students count the letters then say each letter in the term. Have students identify and circle two incidences of the math term. Direct students to trace the word and then draw a picture to show what the term means. Then encourage students to describe their picture to a peer.

58 Grade K • Chapter 6 Subtraction

NAME _____ DATE _____

Lesson 5 Word Web

How Many Are Left?

Trace the math words. Draw a picture in each rectangle that shows the meaning of *are left*.

```
are left
```

Teacher Directions: Provide a description, explanation, or example of the term using images or real objects. Have students say the letters aloud as they trace the math term. Direct students to draw two pictures that represent their math term. Then encourage students to describe their pictures to a peer.

Grade K • Chapter 6 Subtraction 59

NAME _____ DATE _____

Lesson 6 Problem Solving

STRATEGY: Write a Number Sentence

How many **are left**?

Write a Number Sentence

___ ___ ⊖ ___ ___ ⊜ ___ ___

Six minus one equals five.

____ are left.

Teacher Directions: Provide a description, explanation, or example of the boldface terms using images or real objects. Have students count aloud as they put a counter on each butterfly in the picture. Then have them write the number on the first line. Next have students remove a counter for the butterfly that is flying and write that number on the next line. Have students trace the minus sign and equals sign and then say the number sentence. Then have them trace the words in the sentence. Explain the difference between a number sentence and a sentence that uses words. Finally, have students write their answer in the restated question and read the answer sentence aloud.

Grade K • Chapter 6 Subtraction

NAME _____ DATE _____

Lesson 7 Concept Web
Subtract to Take Apart 10

Trace the words and symbols.

- take away
- minus −
- subtract
- equals =
- are left

Draw a picture of one of the words.

Teacher Directions: Review each subtraction term and symbol by providing a description, explanation, or example using images or real objects. Have students say the letters aloud as they trace each math term and symbol. Direct students to draw a picture that represents one of the math terms. Then encourage students to describe their pictures to a peer.

Grade K • Chapter 6 *Subtraction* 61

NAME _____ DATE _____

Chapter 7 Compose and Decompose Numbers 11 to 19

Inquiry of the Essential Question:

How do we show numbers 11 to 19 in another way?

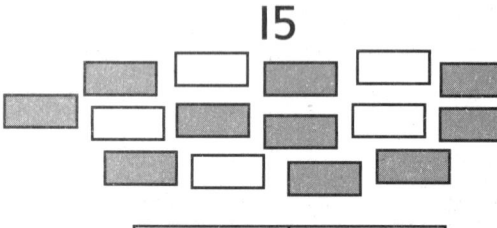

10

7

17

I see ...

I think ...

I know ...

15

10   5

I see ...

I think ...

I know ...

10

4

14

I see ...

I think ...

I know ...

Questions I have...

Teacher Directions: Read the Essential Question for students. Have students echo read. Direct students to describe their observations, inferences, and prior knowledge of each math example. Encourage students to draw additional questions they may have. Scribe questions for students. Then have students share their ideas/questions with a peer.

NAME _____ DATE _____

Lesson 1 Number Identification
Make Numbers 11 to 15

Connect to make a number.

10 and ____ more

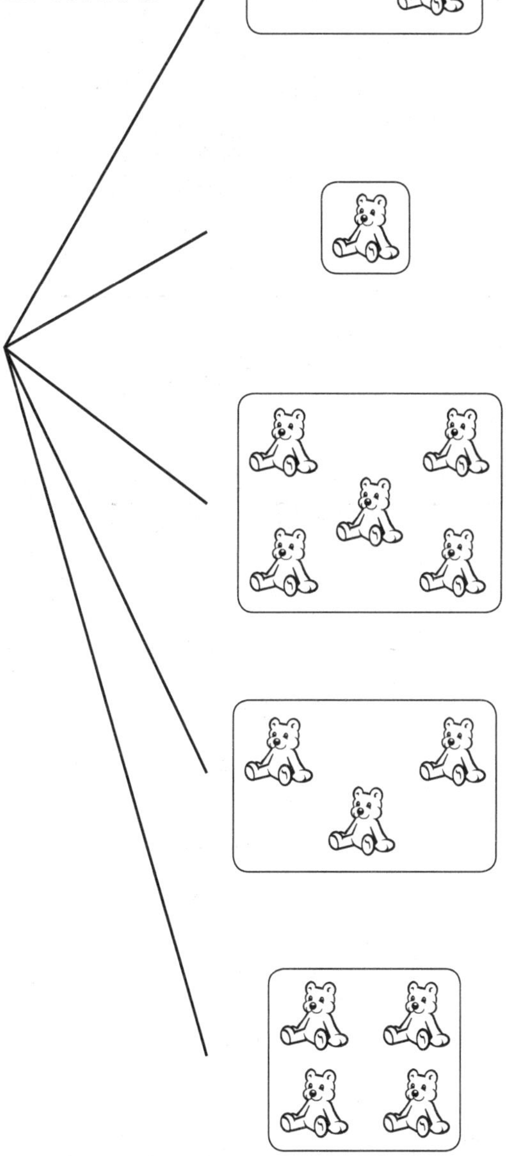

Teacher Directions: Review counting to 15 using images or real objects. Have students say each number then draw a line to connect the two parts of the number. Have partners practice describing each number using a sentence frame such as: **Ten and one more is eleven.**

Grade K • Chapter 7 Compose and Decompose Numbers 11 to 19   63

NAME _____ DATE _____

Lesson 2 Word Web

Take Apart Numbers 11 to 15

Trace the math words. Draw a picture in each rectangle that shows the meaning of *take apart*.

take apart

Teacher Directions: Provide a description, explanation, or example of the new term using images or real objects. Have students say the letters aloud as they trace the math term. Direct students to draw two picture stories that represent their math term. Then encourage students to describe their pictures to a peer.

64 Grade K • Chapter 7 Compose and Decompose Numbers 11 to 19

NAME _____ DATE _____

Lesson 3 Multiple Meaning Word
Problem Solving *STRATEGY: Make a Table*

Say the math word. Trace the word. Draw pictures that show the math word meaning and non-math word meaning in the boxes.

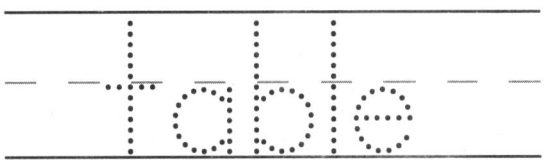

Math Meaning	Non-Math Meaning

A table is _____.

Teacher Directions: Provide math and non-math descriptions, explanations, or examples of the new term using images or real objects. Have students say then trace the term. Then direct students to draw pictures showing a math and non-math meaning of the math term. Model and have students practice the sentence frame: **A table is** _____. Then encourage students to use the sentence frame to describe their pictures to a peer.

Grade K • Chapter 7 *Compose and Decompose Numbers 11 to 19*

Lesson 4 Number Identification
Make Numbers 16 to 19

Trace the numbers. Match.

10 and 7 more is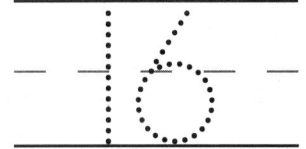

10 and 6 more is 17

10 and 9 more is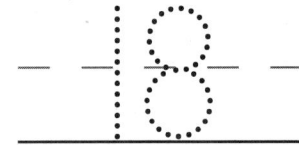

10 and 5 more is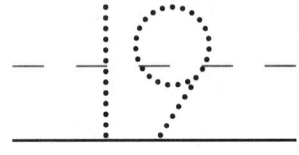

10 and 8 more is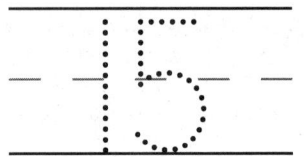

NAME _____ DATE _____

Lesson 5 Number Identification
Take Apart Numbers 16 to 19

Trace the term. Match a number to each picture.

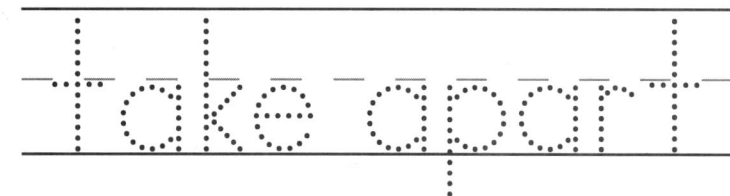

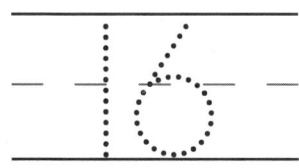
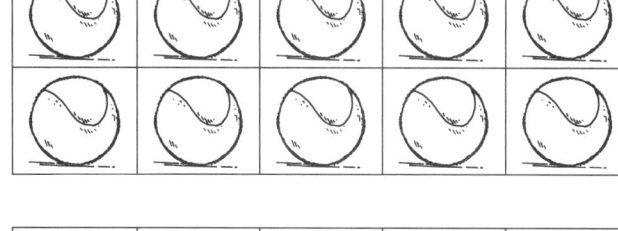

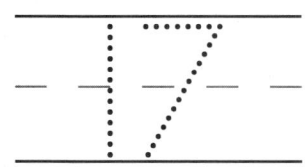
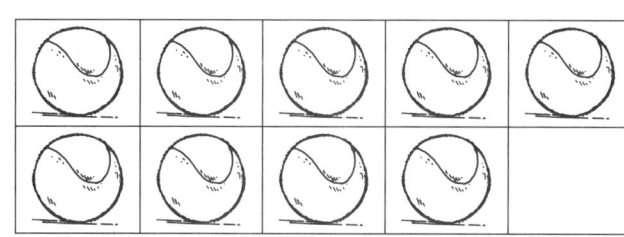

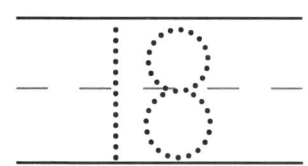
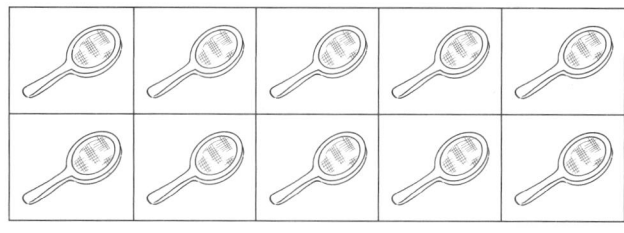

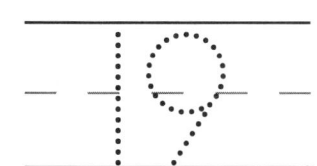
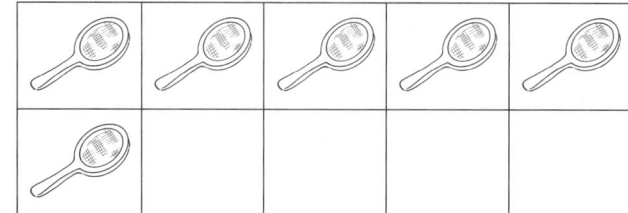

Teacher Directions: Provide a description, explanation, or example of the terms using images or real objects. Have students say the letters aloud as they trace the math term. Direct students to draw a line from a number to a picture that shows a way to take apart that number. Then encourage students to describe each picture to a peer. Suggest sentence frames such as: **I can take apart nineteen. It is ten and nine more.**

Grade K • Chapter 7 *Compose and Decompose Numbers 11 to 19*

NAME _____ DATE _____

Chapter 8 Measurement

Inquiry of the Essential Question:

How do I describe and compare objects by length, height, and weight?

taller shorter

I see ...

I think ...

I know ...

heavier   lighter

I see ...

I think ...

I know ...

shorter

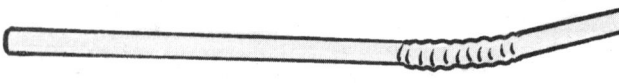

longer

I see ...

I think ...

I know ...

Questions I have...

Teacher Directions: Read the Essential Question for students. Have students echo read. Direct students to describe their observations, inferences, and prior knowledge of each math example. Encourage students to draw additional questions they may have. Scribe questions for students. Then have students share their ideas/questions with a peer.

Lesson 1 Word Web

Compare Length

Trace the math word. Draw a picture in each rectangle that shows the meaning of *length*.

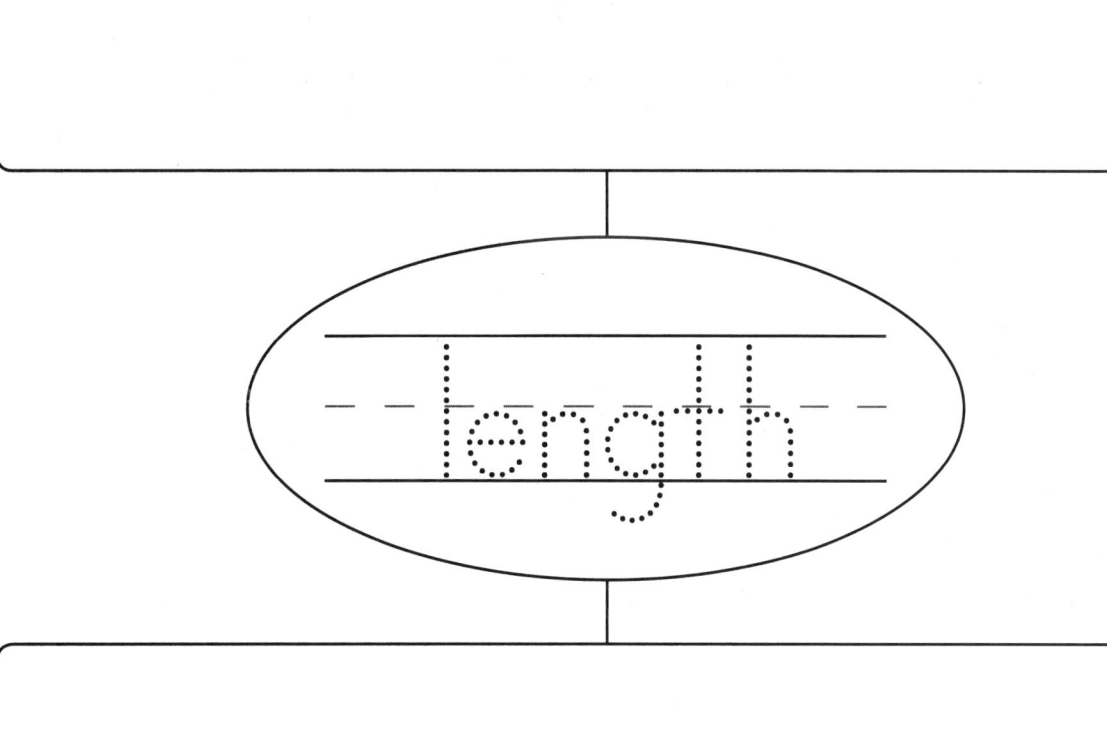

Teacher Directions: Provide a description, explanation, or example of the new term using images or real objects. Discuss and show models for the terms *longer* and *shorter*. Have students say the letters aloud as they trace the math term. Direct students to draw two pictures that represent their math term. Then encourage students to describe their pictures to a peer.

Grade K • Chapter 8 *Measurement*

NAME _____ DATE _____

Lesson 2 Word Journal

Compare Height

Trace the math word *height*. Say the letters as you trace them.

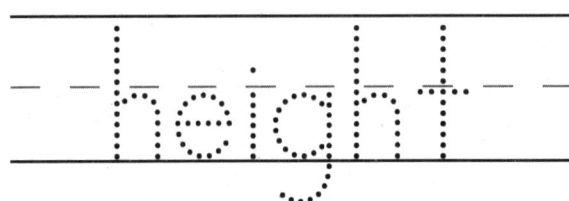

Draw a picture to show what the word *height* means.

Teacher Directions: Provide a description, explanation, or example of the new term using images or real objects. Discuss and show models for the terms *taller* and *shorter* in reference to height. Have students say the letters aloud as they trace the math term. Direct students to draw a picture representing their math term. Then encourage students to describe their picture to a peer.

NAME _____ DATE _____

Lesson 3 Problem Solving

STRATEGY: Guess, Check, and Revise

How **long**?

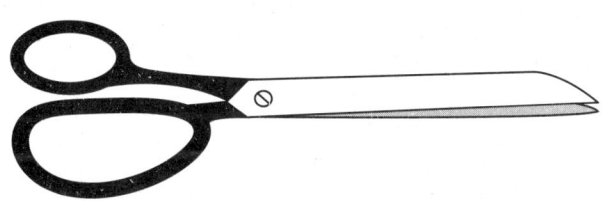

Guess, Check, and Revise

_____   _____

- - - - - - - - -   - - - - - - - - -

_____   _____

         guess   check

It is **about** ____ pennies **long**.

Teacher Directions: Provide a description, explanation, or example of the boldface terms using images or real objects. Review the problem solving strategy using the lesson example. Ask students which item is longer. Model and have students practice the sentences: **This is longer. This is shorter.** Have students tell how many pennies long the shorter object is. **2** Then encourage them to guess how long the longer object is. Finally, have them measure with pennies, write their answer in the restated question, and read the answer sentence aloud.

Grade K • Chapter 8 *Measurement*

NAME _____ DATE _____

Lesson 4 Word Recognition

Compare Weight

Count the letters in the math word *weight* and then say each letter.

		w-e-i-g-h-t			

Circle the math word *weight*.

q	w	w	n	k	i
i	r	e	l	u	v
w	e	i	g	h	t
x	c	g	r	d	e
s	w	h	g	i	h
t	u	t	j	k	r

Trace the word. Draw a picture to show what *weight* means.

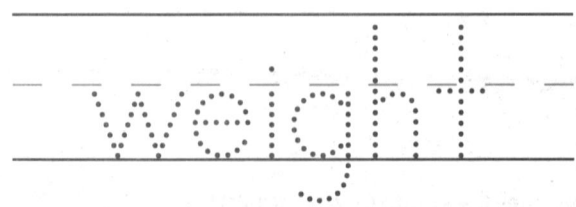

Teacher Directions: Provide a description, explanation, or example of the new term using images or real objects. Discuss and model the terms *heavier* and *lighter* in reference to weight. Have students count the letters and then say each letter in the term. Have students identify, and circle two incidences of the math term. Direct students to trace the word, then draw a picture to show what the term means. Encourage students to describe their picture to a peer.

72 Grade K • Chapter 8 *Measurement*

NAME _____ DATE _____

Lesson 5 Concept Web

Describe Length, Height, and Weight

Trace the word. Look at the circled picture.
Then circle the correct word.

lighter   heavier

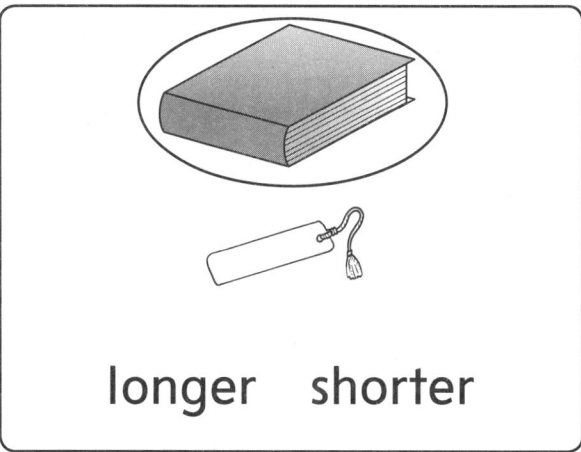

longer   shorter

measure

taller   shorter

Teacher Directions: Review the terms by providing a description, explanation, or example using images or real objects. Direct students to look at the circled picture and then choose the correct word to describe it. Model and have students practice sentences that compare using a sentence frame such as: **The bus is heavier than the butterfly. The butterfly is lighter than the bus.**

Grade K • Chapter 8 *Measurement* **73**

NAME _____ DATE _____

Lesson 6 Vocabulary Word Identification
Compare Capacity

Trace the word. Match.

capacity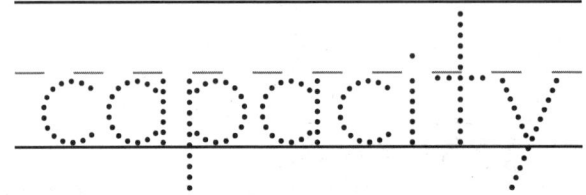

holds more

holds less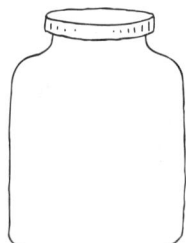

What holds more than a bathtub? Draw a picture of it.

 **Teacher Directions:** Provide a description, explanation, or example of the new term using images or real objects. Have students say each term then draw a line to match each term to a picture. Direct students to draw a picture of something that holds more than a bathtub. Then encourage students to describe their picture to a peer.

NAME _____ DATE _____

Chapter 9 Classify Objects
Inquiry of the Essential Question:

How do I sort objects?

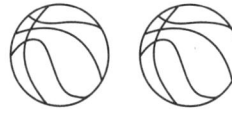

alike

I see ...

I think ...

I know ...

different

I see ...

I think ...

I know ...

I see ...

I think ...

I know ...

Questions I have...

Teacher Directions: Read the Essential Question for students. Have students echo read. Direct students to describe their observations, inferences, and prior knowledge of each math example. Encourage students to draw additional questions they may have. Scribe questions for students. Then have students share their ideas/questions with a peer.

Grade K • Chapter 9 *Classify Objects* 75

Lesson 1 Word Web
Alike and Different

Trace the math words. Draw a picture in each rectangle that shows the meaning of the math word.

[alike]

[different]

Teacher Directions: Provide a description, explanation, or example of the new terms using images or real objects. Have students say the letters aloud as they trace each math term. Direct students to draw pictures that represent each math term. Then encourage students to describe their pictures to a peer.

76 Grade K • Chapter 9 *Classify Objects*

Lesson 2 Problem Solving
STRATEGY: Use Logical Reasoning

What does not **belong**?

Use Logical Reasoning

The _____ does not **belong**.

Teacher Directions: Provide a description, explanation, or example of the boldface term. Point to and teach names of images in the scene. Review the problem solving strategy using the lesson example. Direct students to circle the items that do not belong. Model and practice the sentence frame: **The ____ does not belong.** Have students write an answer in the restated question and read the answer sentence aloud.

Grade K • Chapter 9 *Classify Objects*

NAME _____ DATE _____

Lesson 3 Word Recognition

Sort by Size

Count the letters in the math word *size* then say each letter.

	s-i-z-e

Circle the math word *size*.

n	u	b	e	r	t
v	s	h	p	l	o
m	i	d	w	s	s
e	z	s	i	z	e
z	e	i	j	i	a
x	h	y	q	w	z

Trace the word. Draw a picture to show what *size* means.

Teacher Directions: Provide a description, explanation, or example of the new term using images or real objects. Have students count the letters then say each letter in the term. Have students identify, and circle two incidences of the math term. Direct students to trace the word and then draw a picture to show what the term means. Then encourage students to describe their picture to a peer.

78 Grade K • Chapter 9 *Classify Objects*

NAME _____ DATE _____

Lesson 4 Word Journal

Sort by Shape

Trace the math word *shape*. Say the letters as you trace them.

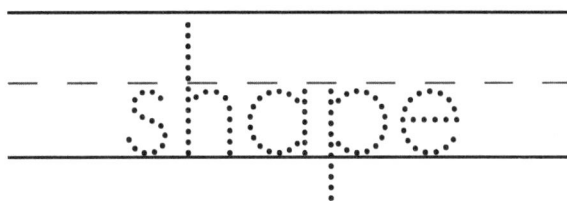

Draw a picture to show what the word means.

Teacher Directions: Provide a description, explanation, or example of the new term using images or real objects. Have students say the letters aloud as they trace the math term. Direct students to draw a picture representing their math term. Then encourage students to describe their picture to a peer.

Grade K • Chapter 9 Classify Objects 79

NAME _____ DATE _____

Lesson 5 Number Identification
Sort by Count

Trace each word. Match.

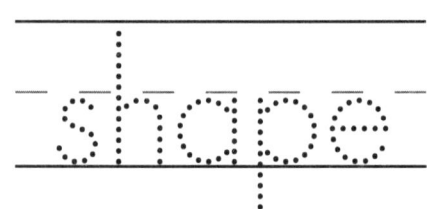 

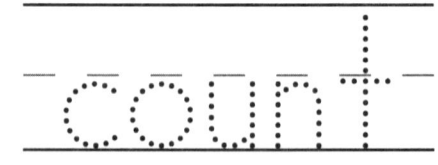 

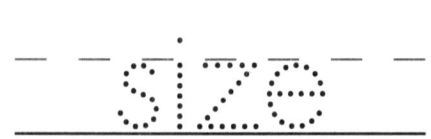 

Draw a picture to show what it means to *sort by count*.

Teacher Directions: Review the meanings of each word by providing a description, explanation, or example of the terms using images or real objects. Have students say the letters aloud as they trace each math term. Direct students to draw a picture representing how to sort by count. Then encourage students to describe their picture to a peer.

80 Grade K • Chapter 9 *Classify Objects*

NAME _____ DATE _____

Chapter 10 Position

Inquiry of the Essential Question:

How do I identify positions?

behind

I see ...

I think ...

I know ...

below

I see ...

I think ...

I know ...

beside

I see ...

I think ...

I know ...

Questions I have...

Teacher Directions: Read the Essential Question for students. Have students echo read. Direct students to describe their observations, inferences, and prior knowledge of each math example. Encourage students to draw additional questions they may have. Scribe questions for students. Then have students share their ideas/questions with a peer.

Grade K • Chapter 10 *Position*

Lesson 1 Vocabulary Word Identification
Above and Below

Circle *above* and *below*.

below	alike	long	shape
above	five	different	below
size	below	above	eight
yellow	sort	short	below

Draw a picture of your math words.

Teacher Directions: Provide a description, explanation, or example of the new terms using images or real objects. Have students identify, and circle each incidence of each math term. Direct students to draw a picture representing their math terms. Then encourage students to describe their picture to a peer.

Lesson 2 Word Recognition
In Front of and Behind

Count the letters in the math word *behind* then say each letter.

b-e-h-i-n-d

Circle the math word *behind*.

p	b	u	m	i	n
b	e	h	i	n	d
e	h	x	o	n	g
t	i	a	q	f	c
s	n	d	e	i	n
j	d	x	e	t	n

Trace the words. Draw a picture to show what *in front of* means.

 **Teacher Directions:** Provide a description, explanation, or example of the new terms using images or real objects. Have students count the letters then say each letter in the term. Have students identify and circle two incidences of the math term *beside* in the box. Direct students to trace the term *in front of*, then draw a picture to show what the term means. Encourage students to describe their picture to a peer.

Grade K • Chapter 10 Position 83

NAME _____ DATE _____

Lesson 3 Word Web

Next to and Beside

Trace the math words. Draw a picture in each rectangle that shows the meaning of *next to* and *beside*.

next to

beside

Teacher Directions: Provide a description, explanation, or example of the new terms using images or real objects. Explain that *next to* and *beside* have the same meaning. Model and practice a sentence frame such as: **The cat is next to the dog.** Then practice the same sentence using *beside*. Have students say the letters aloud as they trace each math term. Direct students to draw two pictures that represent their math terms. Then encourage students to describe their pictures to a peer.

84 Grade K • Chapter 10 *Position*

Lesson 4 Problem Solving
STRATEGY: Act It Out

Where do I **put** it?

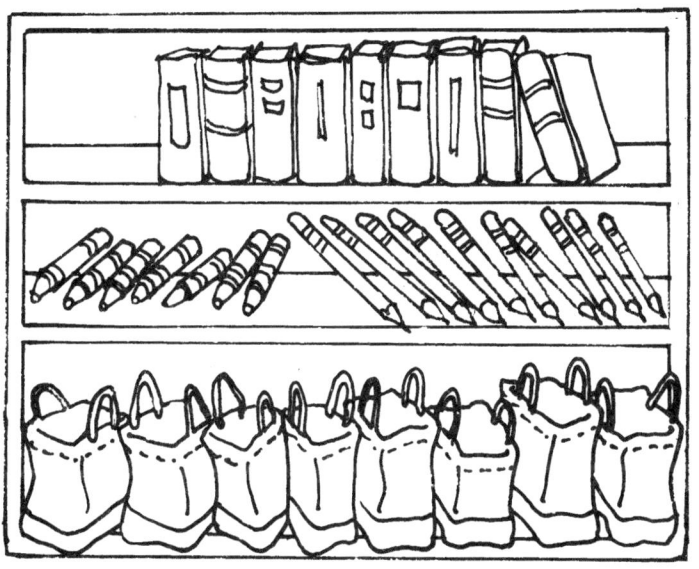

Act It Out

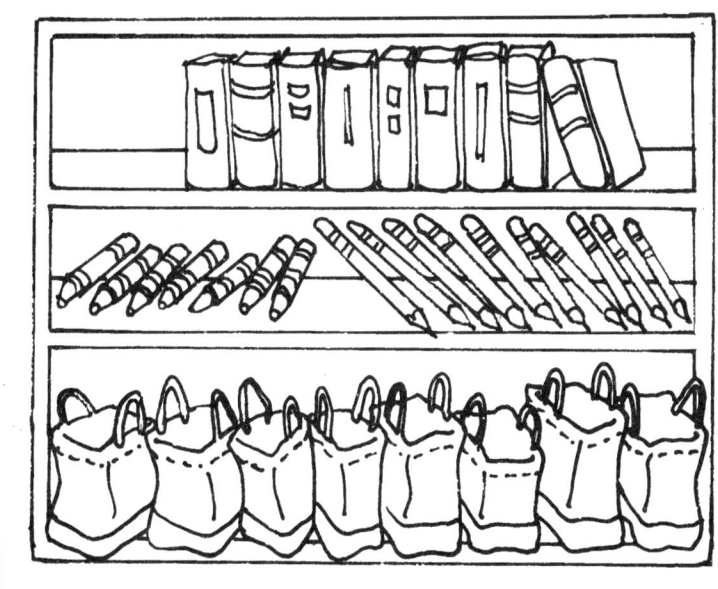

I can **put** it _____ the pencils.

Teacher Directions: Provide a description, explanation, or example of the boldface terms using gestures. Review the problem solving strategy using the lesson example. Direct students to use a connecting cube to show where the book should be placed in the top picture. Have them draw the book where it belongs in the bottom picture. Have students write their answer in the restated question and read the answer sentence aloud.

NAME _____ DATE _____

Chapter 11 Two-Dimensional Shapes
Inquiry of the Essential Question:

How can I compare shapes?

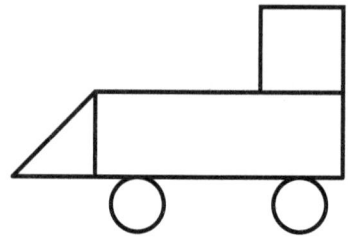

I see ...

I think ...

I know ...

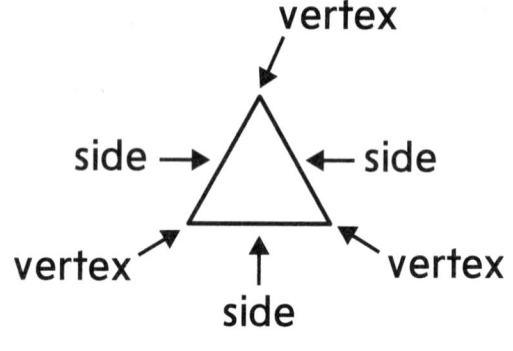

I see ...

I think ...

I know ...

 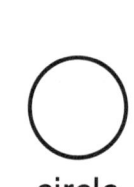 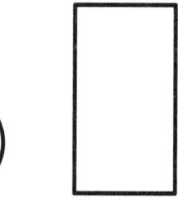  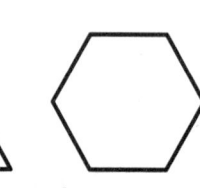
square   circle   rectangle   triangle   hexagon

I see ...

I think ...

I know ...

Questions I have...

Teacher Directions: Read the Essential Question for students. Have students echo read. Direct students to describe their observations, inferences, and prior knowledge of each math example. Encourage students to draw additional questions they may have. Scribe questions for students. Then have students share their ideas/questions with a peer.

NAME _____ DATE _____

Lesson 1 Word Web
Squares and Rectangles

Label each item with a word from the word bank.

Word Bank
rectangle   side   square   vertex

two-dimensional shapes

Teacher Directions: Provide a description, explanation, or example of the new terms using images or real objects. Teach students that the plural of *vertex* is *vertices*. Direct students to look at each picture and then label it with the correct word from the word bank. Model and have students practice sentences that describe each shape, such as: **A square has four sides and four vertices.**

Grade K • Chapter 11 *Two-Dimensional Shapes* **87**

NAME _____ DATE _____

Lesson 2 Word Recognition
Circles and Triangles

Count the letters in the math word *round* then say each letter.

r-o-u-n-d

Circle the math word *round*.

a	r	o	u	n	d
g	r	h	p	l	o
l	o	t	g	r	c
e	u	s	x	a	z
p	n	n	y	h	t
f	d	d	e	b	h

Trace the word. Draw a picture to show what *straight* means.

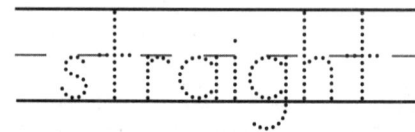

Teacher Directions: Provide a description, explanation, or example of the new terms using images or real objects. Have students count the letters then say each letter in *round*. Have students identify, and circle two incidences of the math term. Direct students to trace the word *straight*, then draw a picture to show what the term means. Encourage students to describe their picture to a peer. If necessary, provide a sentence frame such as: **The sides of a triangle are straight.**

88 Grade K • Chapter 11 *Two-Dimensional Shapes*

NAME _____ DATE _____

Lesson 3 Shape Identification
Squares, Rectangles, Triangles, and Circles

Match.

 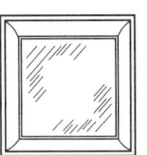  

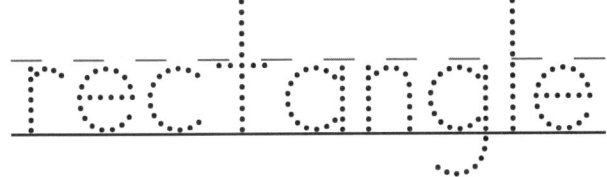   

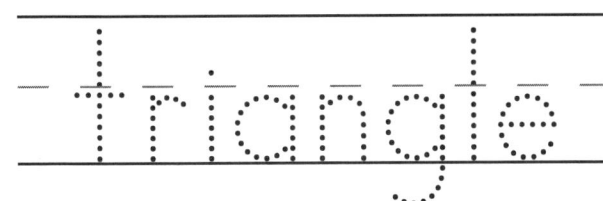   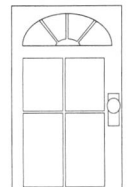

Draw all the shapes. Say the name of each shape.

Teacher Directions: Review the names of each shape using images or real objects. Have students say each shape then draw a line to match to a picture. Direct students to draw the shapes and say their names. Then encourage students to describe their picture to a peer. Encourage them to use the shape names and the terms *side, vertex (vertices), round,* and *straight* in their descriptions.

Grade K • Chapter 11 *Two-Dimensional Shapes* **89**

NAME _____ DATE _____

Lesson 4 Word Journal
Hexagons

Trace the math word *hexagon*. Say the letters as you trace them.

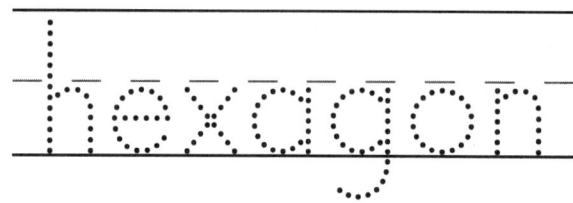

Draw a picture to show what *hexagon* means.

Teacher Directions: Provide a description, explanation, or example of the new term using images or real objects. Have students say the letters aloud as they trace the math term. Direct students to draw a picture representing their math term. Then encourage students to describe their picture to a peer. Encourage them to count then tell how many sides and vertices the shape has.

90 Grade K • Chapter 11 *Two-Dimensional Shapes*

Lesson 5 Vocabulary Word Identification
Shapes and Patterns

Circle the math word *pattern* in the boxes below.

square	circle	ten	pattern
more	long	pattern	alike
count	pattern	shape	size
pattern	round	side	vertex

Draw a picture of your math word.

Teacher Directions: Provide a description, explanation, or example of the new term using images or real objects. Have students identify, and circle each incidence of the math term. Direct students to draw a picture representing their math term. Then encourage students to describe their picture to a peer.

NAME _____ DATE _____

Lesson 6 Concept Web
Shapes and Position

Trace the word. Draw a sun ☼ in the correct position.

- above
- behind
- next to
- position
- below
- beside
- in front of

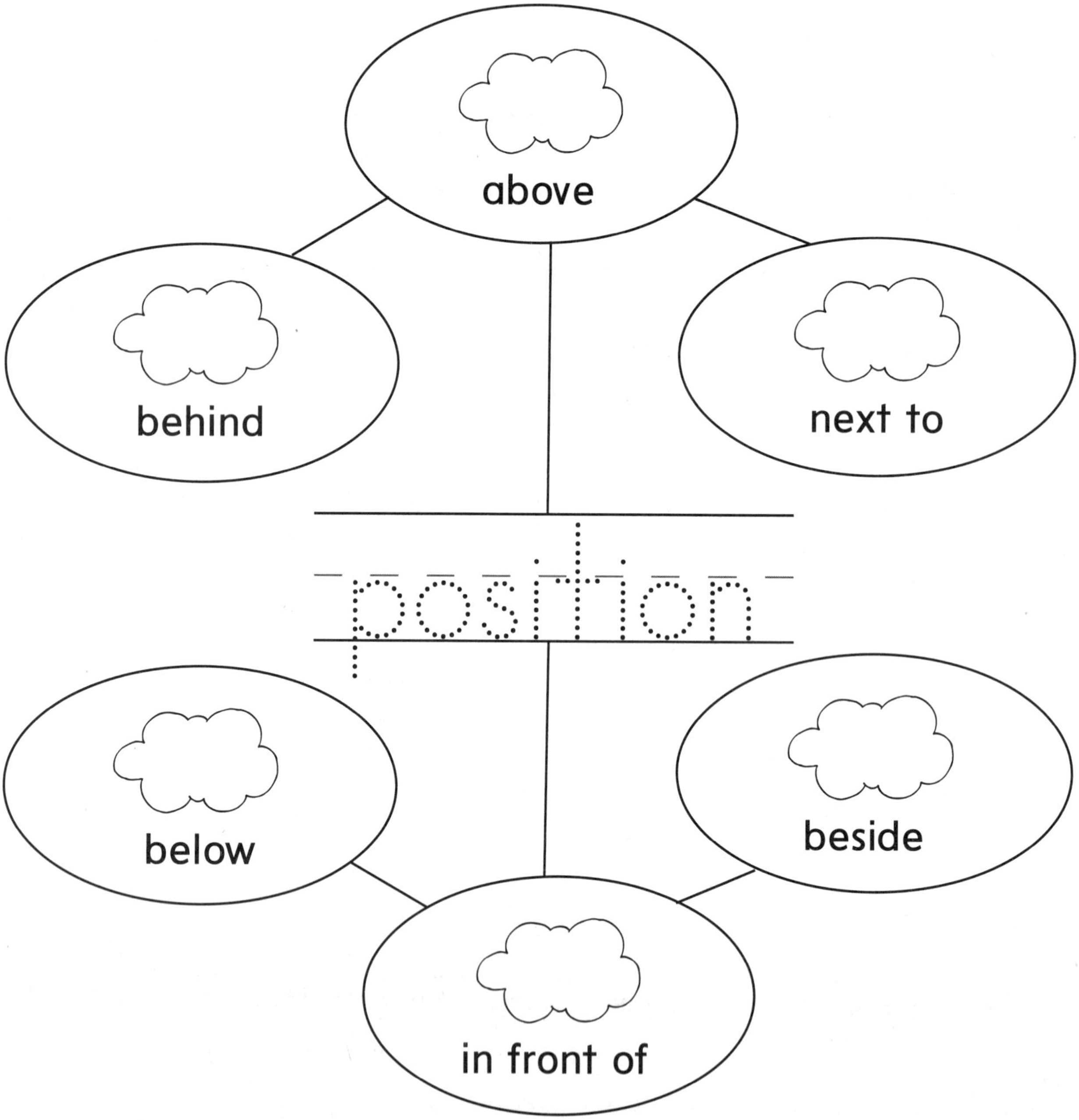

Teacher Directions: Provide a description, explanation, or example of the terms using images or real objects. Have students trace the word, saying each letter as they write it. Then have them draw a sun in the correct position in relation to the cloud. For example, for *above*, students should draw a sun above the cloud. Model and prompt students to practice sentences that tell position, such as: **The sun is above the cloud.**

92 Grade K • Chapter 11 *Two-Dimensional Shapes*

NAME _____ DATE _____

Lesson 7 Word Web
Compose New Shapes

Trace the math word. Draw a picture in each rectangle that shows the meaning of the math word.

shapes

Teacher Directions: Provide a description, explanation, or example of the term using images or real objects. Have students say the letters aloud as they trace the math term. Direct students to draw two pictures that represent their math term. Then encourage students to describe their pictures to a peer.

Grade K • Chapter 11 *Two-Dimensional Shapes* 93

NAME _____ DATE _____

Lesson 8 Problem Solving
STRATEGY: Use Logical Reasoning

What shapes are **missing**?

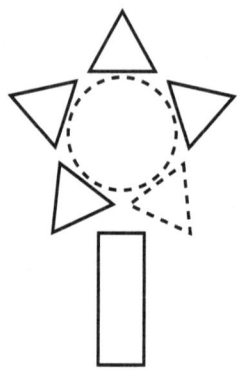

Use Logical Reasoning

A _____ and a _____ are **missing**.

Teacher Directions: Provide a description, explanation, or example of the bold face term using images or real objects. Review the problem solving strategy using the lesson example. Direct students to identify the missing shapes and draw them. Have students write their answer in the restated question and read the answer sentence aloud.

94 Grade K • Chapter 11 *Two-Dimensional Shapes*

NAME _____ DATE _____

Lesson 9 Word and Shape Identification
Model Shapes in the World

Fill in the missing words and numbers. Then draw the shape.

Word	Numbers	Picture
circle	____ sides ____ vertices	
hexagon	____ sides ____ vertices	
rectangle	____ sides ____ vertices	
square	____ sides ____ vertices	
triangle	____ sides ____ vertices	

Teacher Directions: Use attribute blocks to review the name of each shape. Have students point out the vertices and sides. For each item, have students trace the word. Then have students fill in the missing number of sides and vertices for each shape. Encourage partners to report about a shape using a sentence frame such as: **A triangle has 3 sides and 3 vertices.** For the circle, model and prompt students how to report 0 sides and vertices: **A circle has zero sides. A circle has no vertices.**

Grade K • Chapter 11 *Two-Dimensional Shapes* 95

NAME _____ DATE _____

Chapter 12 Three-Dimensional Shapes
Inquiry of the Essential Question:

How do I identify and compare three-dimensional shapes?

I see ...

I think ...

I know ...

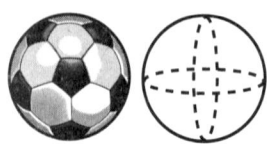

I see ...

I think ...

I know ...

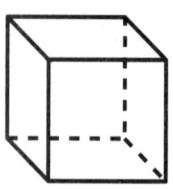

I see ...

I think ...

I know ...

Questions I have...

Teacher Directions: Read the Essential Question for students. Have students echo read. Direct students to describe their observations, inferences, and prior knowledge of each math example. Encourage students to draw additional questions they may have. Scribe questions for students. Then have students share their ideas/questions with a peer.

NAME _____ DATE _____

Lesson 1 Concept Web
Spheres and Cubes

Look at the shape. Circle the correct word.

sphere cube	spheres cubes

sphere 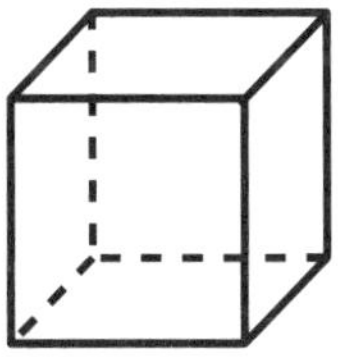   cube

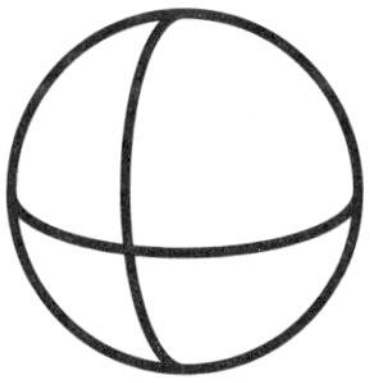

spheres cubes	sphere cube

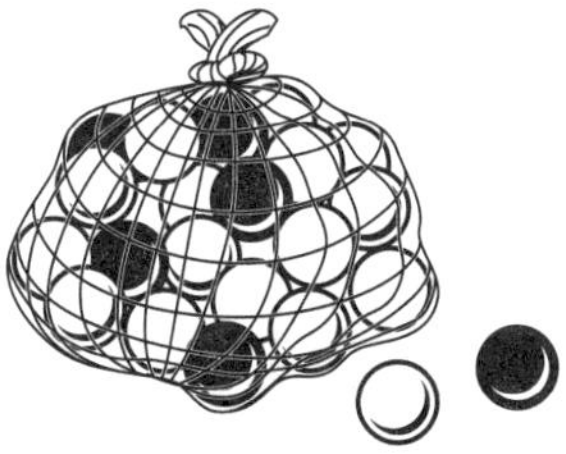

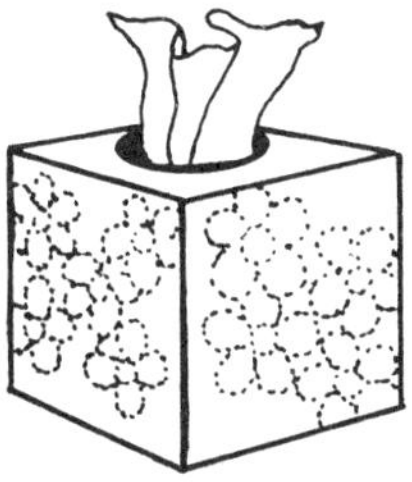

 **Teacher Directions:** Review the terms by providing a description, explanation, or example using images or real objects. Allow extra time for students to practice the difficult /sf/ sound at the beginning of *sphere*. Teach the plural form of *sphere (spheres)* and *cube (cubes)*. Direct students to look at the picture and then circle the correct word to name it. Model and have students practice sentences such as: **The soccer ball is a sphere. The marbles are spheres.**

Grade K • Chapter 12 *Three-Dimensional Shapes* **97**

NAME _____ DATE _____

Lesson 2 Concept Web
Cylinders and Cones

Look at the shape. Draw a line to the correct word.

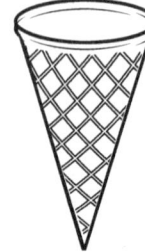 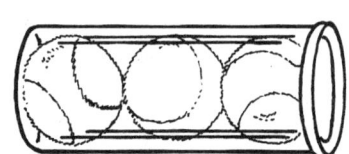  

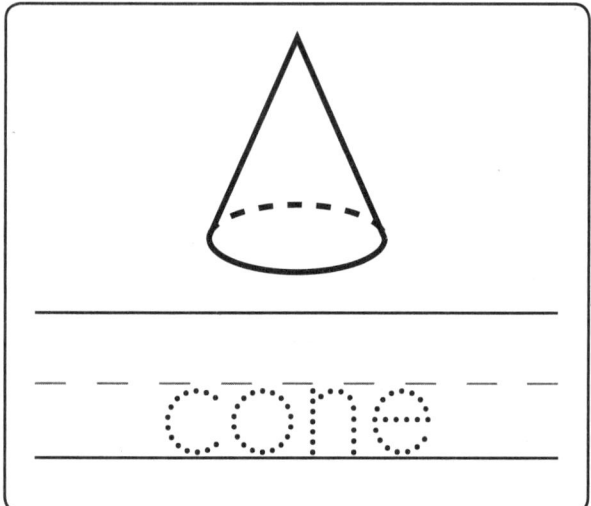

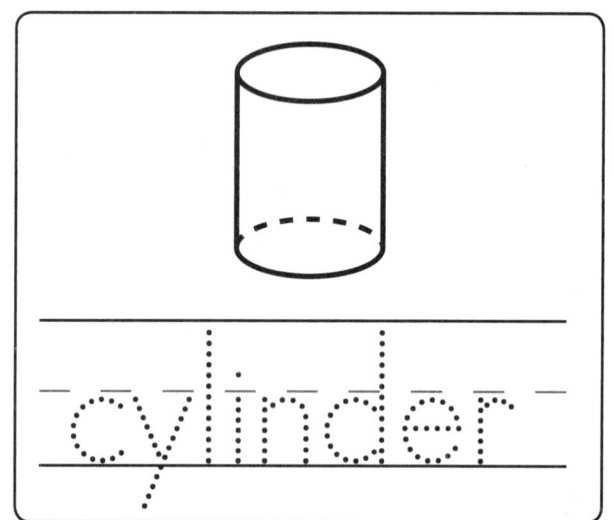

   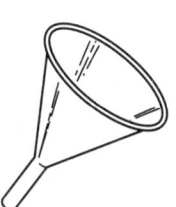

 **Teacher Directions:** Review the terms by providing a description, explanation, or example using images or real objects. Teach the names of the pictured objects. Direct students to look at the picture and then draw a line from the picture to the correct word to match its shape. Model and have students practice sentences such as: **The party hat is a cone. The bucket is a cylinder.**

Lesson 3 Word and Shape Identification
Compare Solid Shapes

Trace each word. Match.

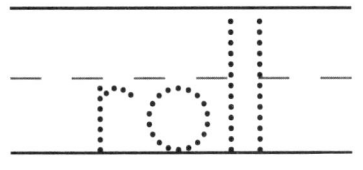  

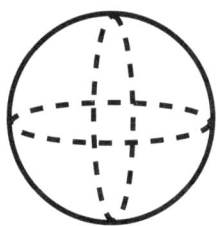   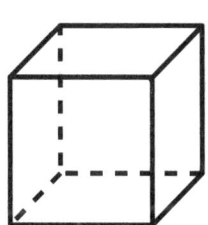

Draw objects that roll.

Teacher Directions: Review the solid shapes using images or real objects. Then use the shapes to teach roll, stack, and slide. Have students say the name of each shape and then draw a line to match each shape to a word. Some shapes will match to more than one word. Direct students to draw objects that roll and name them. Then encourage students to describe their picture to a peer.

Grade K • Chapter 12 *Three-Dimensional Shapes*

NAME _____ DATE _____

Lesson 4 Problem Solving
STRATEGY: Act It Out

What **will stack on** the **cube**?

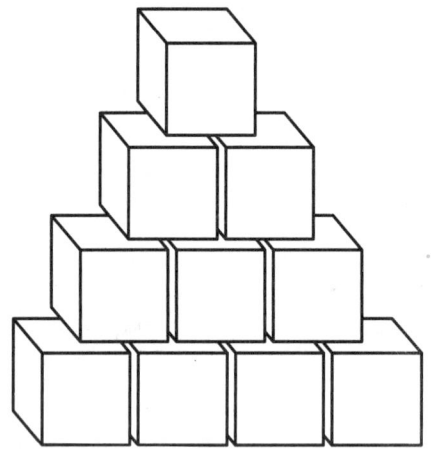

Act It Out

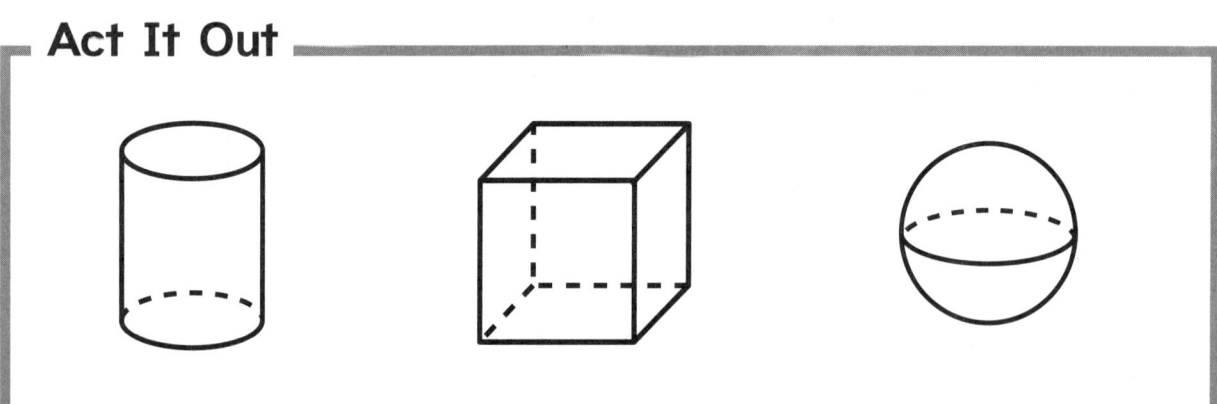

The _____ and the

_____ **will stack on** the **cube**.

Teacher Directions: Provide a description, explanation, or example of the boldface terms using images or real objects. Review the problem solving strategy using the lesson example. Direct students to use solid shapes to act out the problem. Have them circle the items that will stack on the cube. Then have them cross out the item that will not stack. Have students write their answer in the restated question and read the answer sentence aloud.

100 Grade K • Chapter 12 *Three-Dimensional Shapes*

NAME _____ DATE _____

Lesson 5 Word and Shape Identification
Model Solid Shapes in Our World

Look at the picture. Then complete the sentence.

Word Bank			
cone	cube	cylinder	sphere

Shape	Sentence
(block with letter A)	I am a _____.
(party hat with stars and moons)	I am a _____.
(baseball)	I am a _____.
(candle)	I am a _____.

Teacher Directions: Review the name of each shape. Have students tell which shapes will roll, stack, or slide. For each item, have students look at the "real-world" picture and determine which shape it is. Then have students use a word from the word bank to complete each sentence. Encourage them to read their sentences to a peer.

What are VKVs® and How Do I Create Them?

Visual Kinesthetic Vocabulary Cards® are flashcards that animate words by focusing on their structure, use, and meaning. The VKVs in this book are used to show cognates, or words that are similar in Spanish and English.

Step 1
Go to the back of your book to find the VKVs for the chapter vocabulary you are currently studying. Follow the cutting and folding instructions at the top of the page. The vocabulary word on the BLUE background is written in English. The Spanish word is on the ORANGE background.

Step 2
There are exercises for you to complete on the VKVs. When you understand the concept, you can complete each exercise. All exercises are written in English and Spanish. You only need to give the answer once.

Step 3
Individualize your VKV by writing notes, sketching diagrams, recording examples, and forming plurals.

How Do I Store My VKVs?
Take a 6" x 9" envelope and cut away a V on one side only. Glue the envelope into the back cover of your book. Your VKVs can be stored in this pocket!

Remember you can use your VKVs ANY time in the school year to review new words in math, and add new information you learn. Why not create your own VKVs for other words you see and share them with others!

Visual Kinesthetic Learning **VKVi**

¿Qué son las VKV y cómo se crean?

Las tarjetas de vocabulario visual y cinético (VKV) contienen palabras con animación que está basada en la estructura, uso y significado de las palabras. Las tarjetas de este libro sirven para mostrar cognados, que son palabras similares en español y en inglés.

Paso 1
Busca al final del libro las VKV que tienen el vocabulario del capítulo que estás estudiando. Sigue las instrucciones de cortar y doblar que se muestran al principio. La palabra de vocabulario con fondo AZUL está en inglés. La de español tiene fondo NARANJA.

Paso 2
Hay ejercicios para que completes con las VKV. Cuando entiendas el concepto, puedes completar cada ejercicio. Todos los ejercicios están escritos en inglés y español. Solo tienes que dar la respuesta una vez.

Paso 3
Da tu toque personal a las VKV escribiendo notas, haciendo diagramas, grabando ejemplos y formando plurales.

¿Cómo guardo mis VKV?
Corta en forma de "V" el lado de un sobre de 6" X 9". Pega el sobre en la contraportada de tu libro. Puedes guardar tus VKV en esos bolsillos. ¡Así de fácil!

Recuerda que puedes usar tus VKV en cualquier momento del año escolar para repasar nuevas palabras de matemáticas, y para añadir la nueva información. También puedes crear más VKV para otras palabras que veas, y poder compartirlas con los demás.

VKV2 Visual Kinesthetic Learning

Chapter 1

✂ cut on all dashed lines   ⬜ fold on all solid lines

3 2 5 0

number

1

one

2

two

Chapter 1 Visual Kinesthetic Learning   VKV3

Chapter 1

cut on all dashed lines   fold on all solid lines

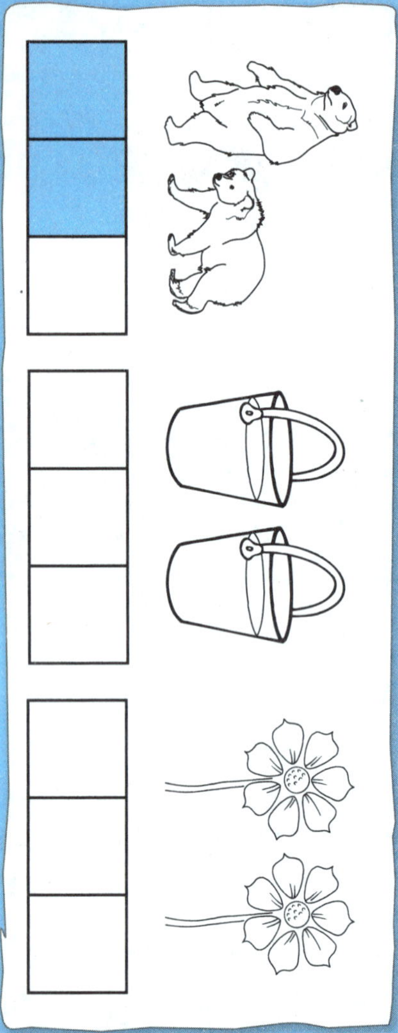

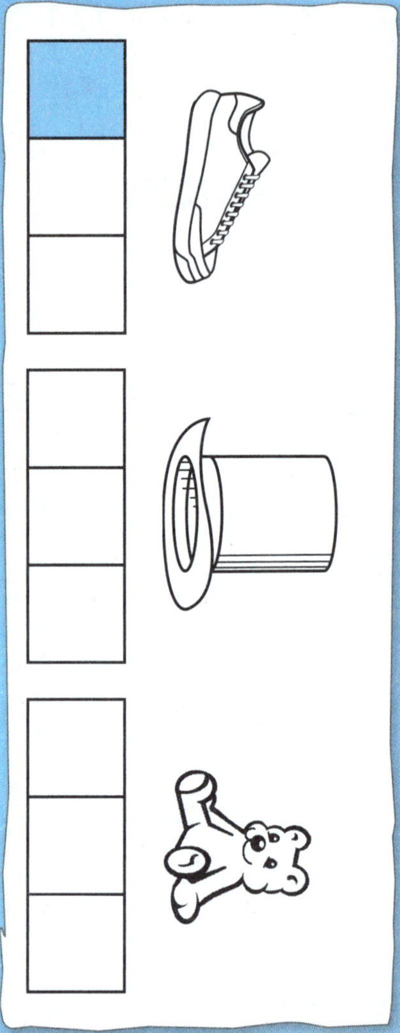

número

uno

dos

one

VKV4 Chapter 1 Visual Kinesthetic Learning

Chapter 1

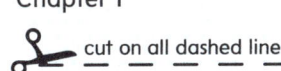

cut on all dashed lines   fold on all solid lines

3

three

mayor

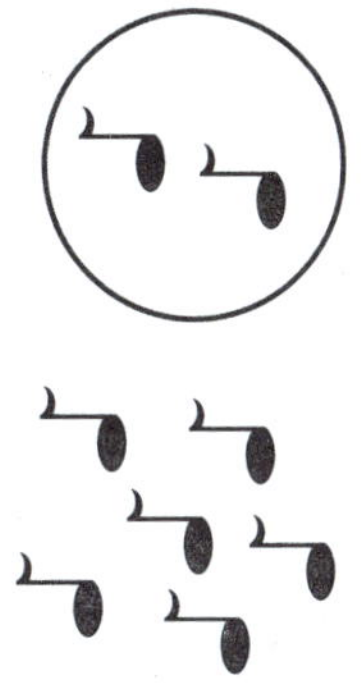

less than

Chapter 1 Visual Kinesthetic Learning   VKV5

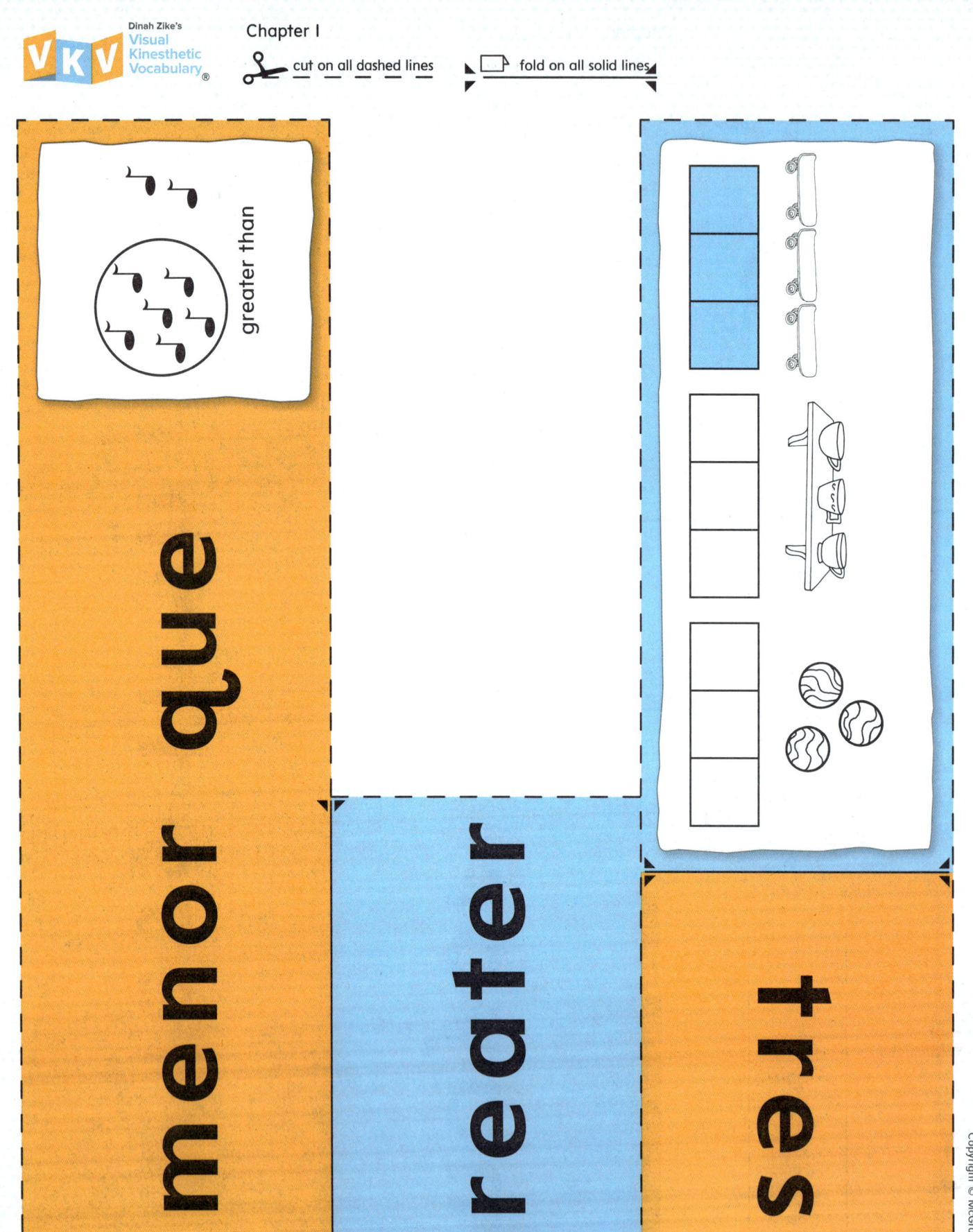

Chapter 2

cut on all dashed lines   fold on all solid lines

ordinal number

first
second
third
fourth

Chapter 2 Visual Kinesthetic Learning   VKV7

Chapter 2

cut on all dashed lines   fold on all solid lines

número ordinal

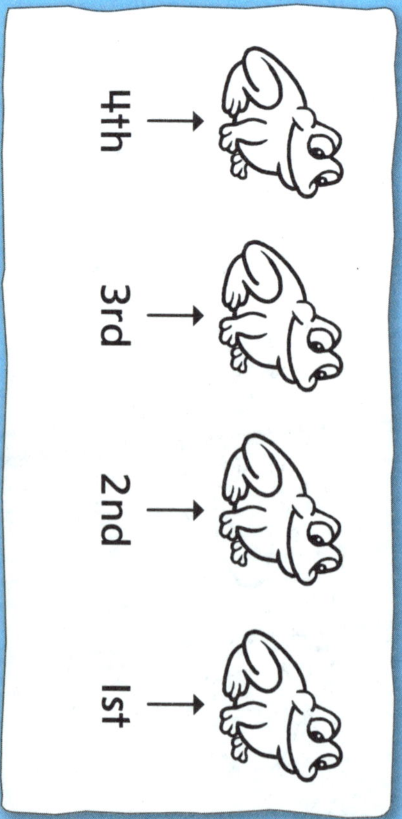

4th →
3rd →
2nd →
1st →

VKV8 Chapter 2 Visual Kinesthetic Learning

Chapter 3

✂ cut on all dashed lines   ▭ fold on all solid lines

15

quince

four   13   thir

14   15   13

Chapter 3 Visual Kinesthetic Learning VKV9

Chapter 3

cut on all dashed lines   fold on all solid lines

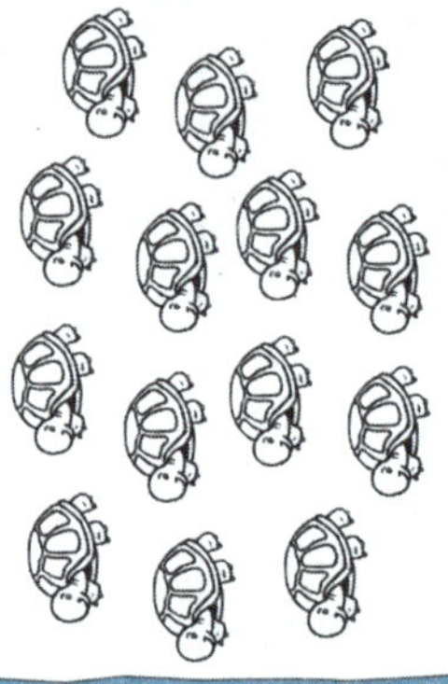

cator   fifteen   tre

14   15   13

VKV10   Chapter 3 Visual Kinesthetic Learning

Chapter 3

cut on all dashed lines   fold on all solid lines

sixteen

siete

nueve

ocho

Chapter 3 Visual Kinesthetic Learning **VKVII**

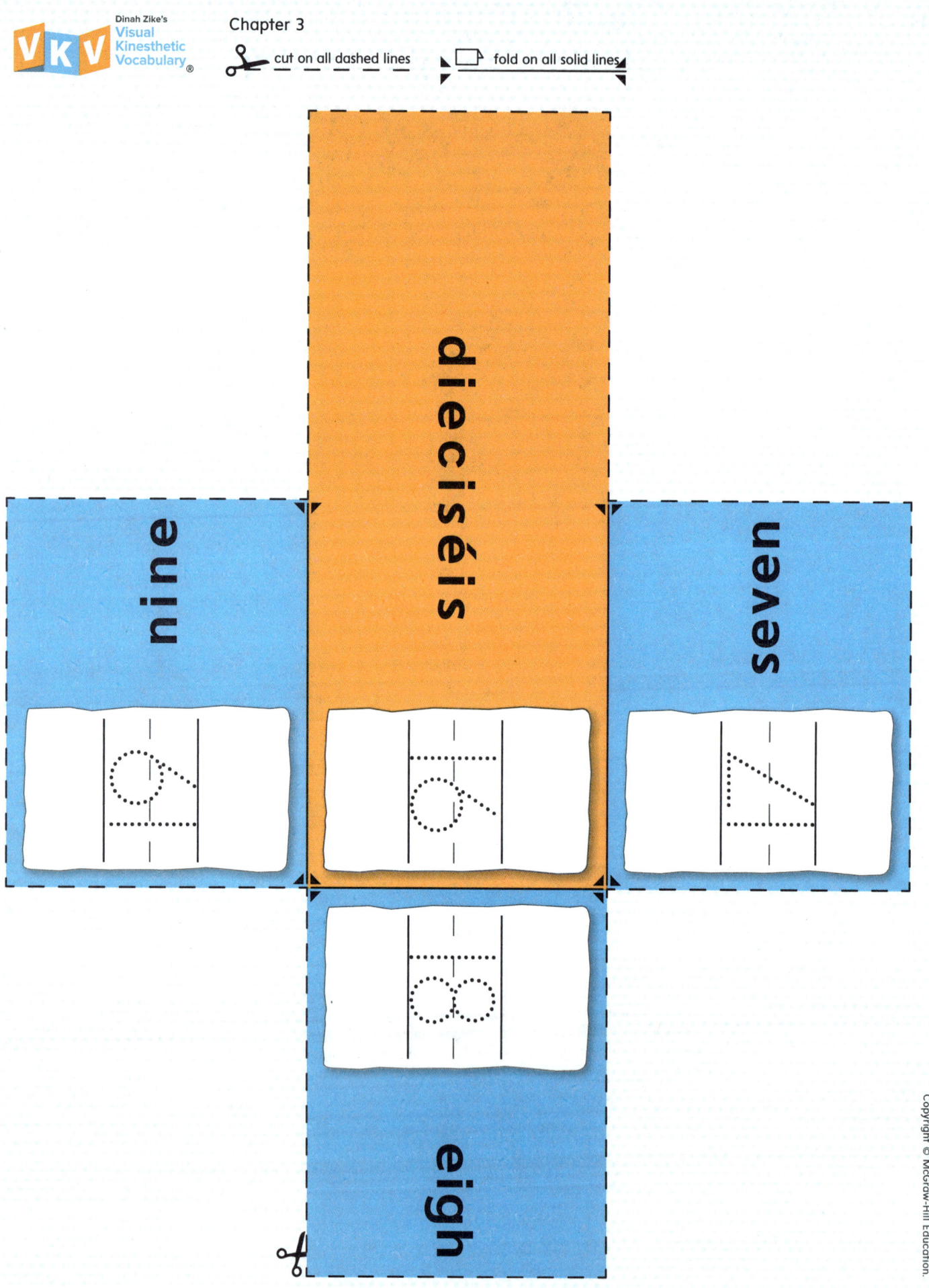

Chapter 5

✂ cut on all dashed lines   fold on all solid lines

addition

join

Chapter 5

suma adición

___ in all

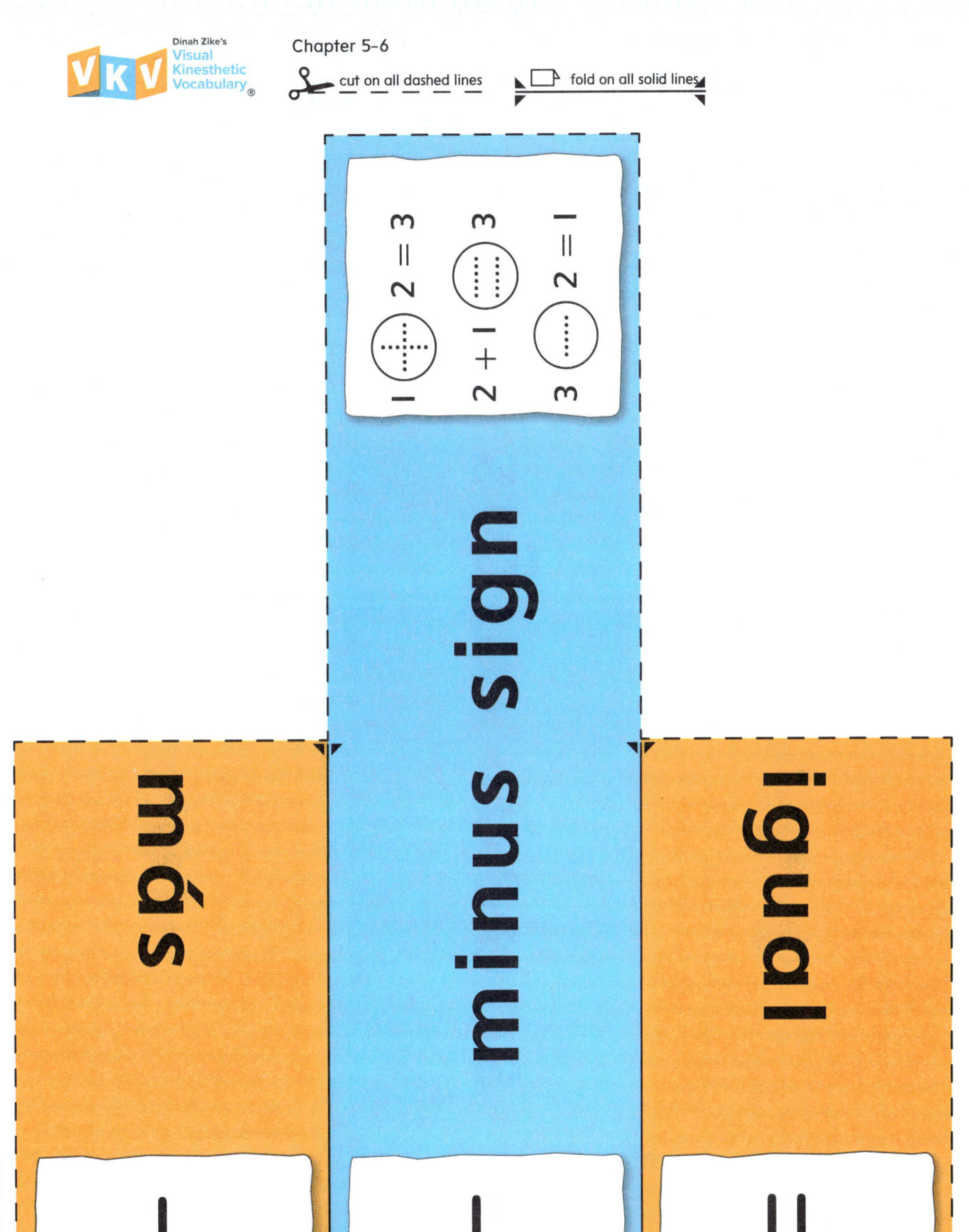

Chapter 5-6

✂ cut on all dashed lines   ▱ fold on all solid lines

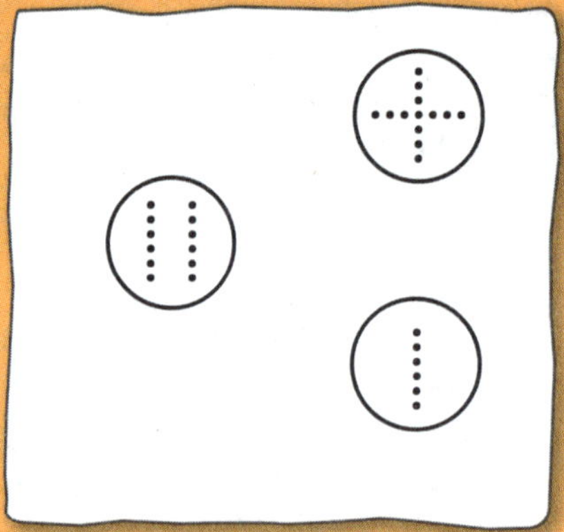

signo menos

equal

plus

=   −   +

Chapter 6

cut on all dashed lines   fold on all solid lines

subtract

Chapter 6

resta | ion

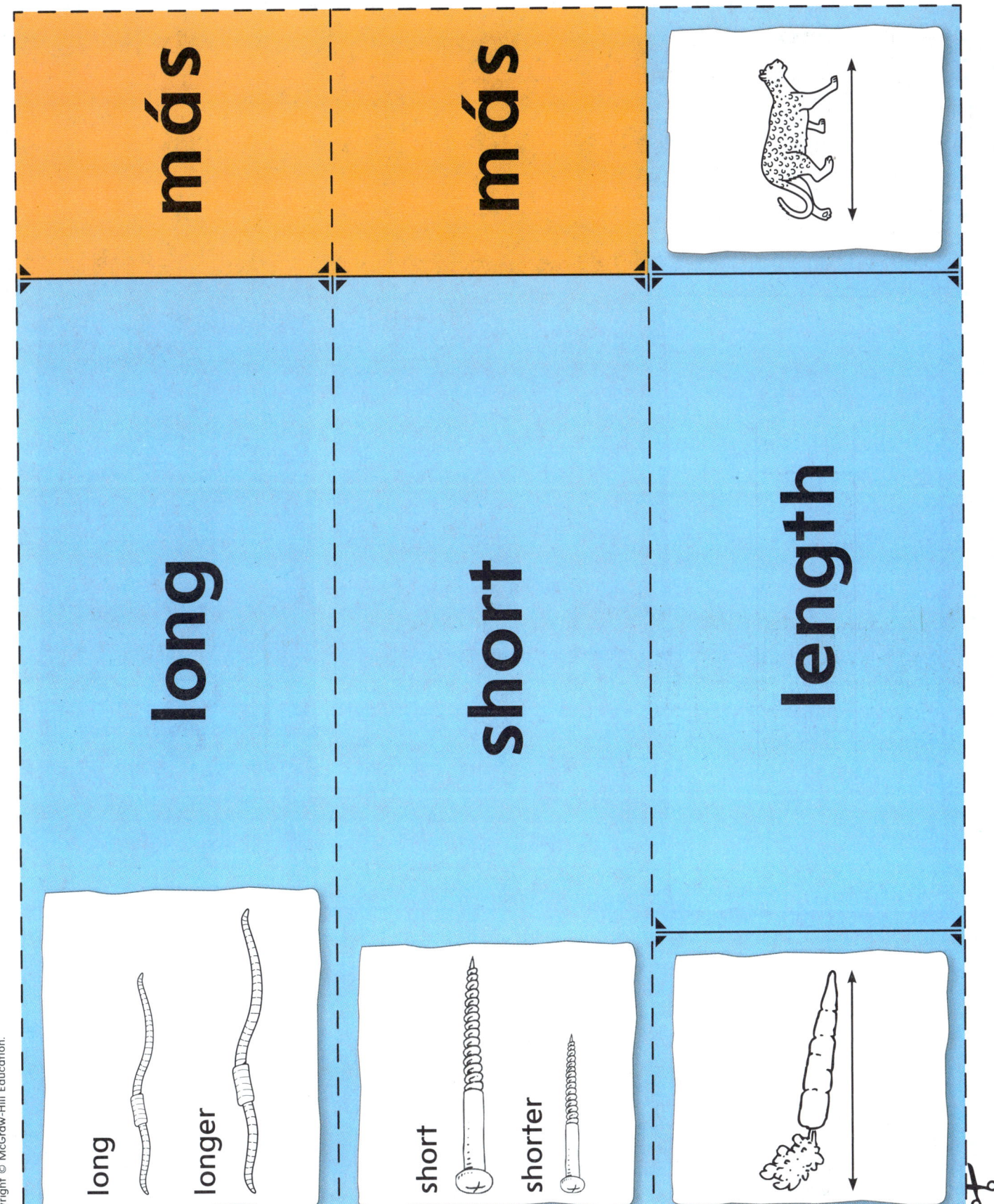

Chapter 8

✂ cut on all dashed lines   📄 fold on all solid lines

itud	er	er
length	corto	largo
lo	Which is shorter? 🍴🍴	Which is longer? 🔨🔨

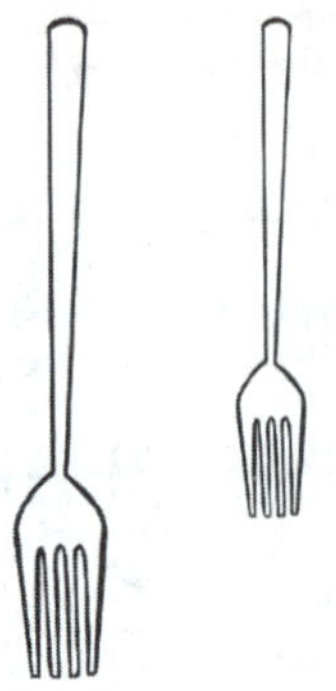

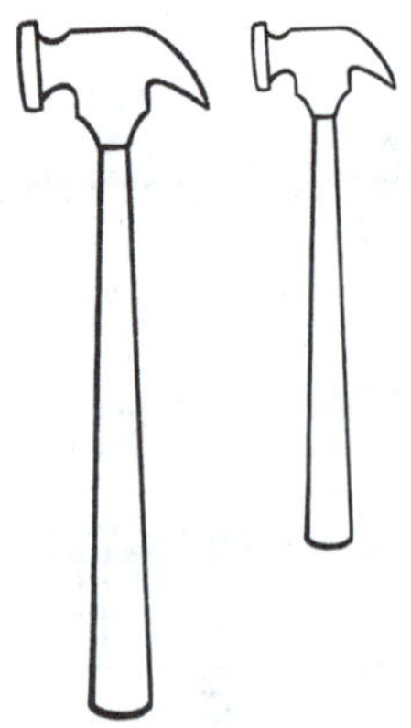

VKV20 Chapter 8 Visual Kinesthetic Learning

Copyright © McGraw-Hill Education.

Chapter 8

cut on all dashed lines fold on all solid lines

más	más	más
tall	light	heavy
taller / tall	lighter / light	heavier / heavy

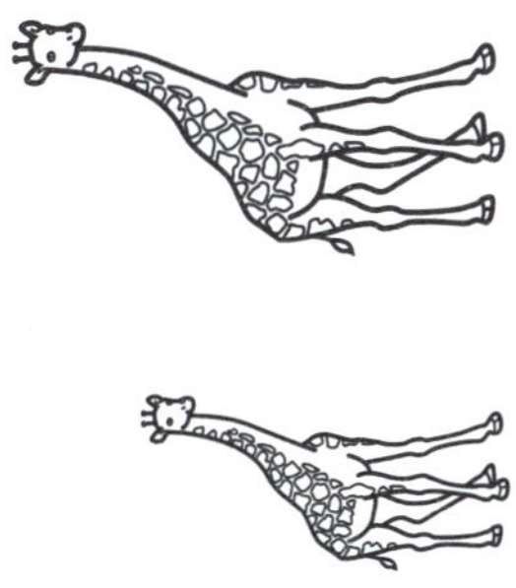

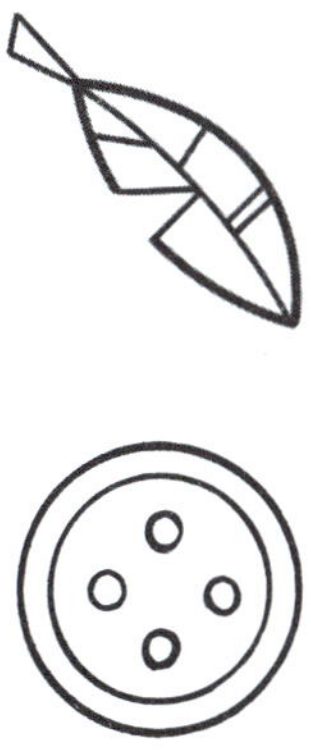

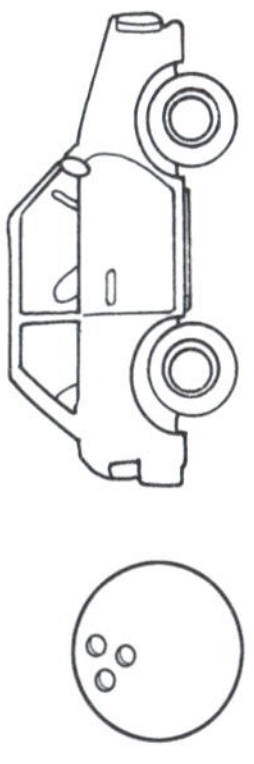

Chapter 8 Visual Kinesthetic Learning VKV21

Chapter 8
- cut on all dashed lines
- fold on all solid lines

ier	er	er
pesado	liviano	alto
Which is heavier?	Which is lighter?	Which is taller?

VKV22 Chapter 8 Visual Kinesthetic Learning

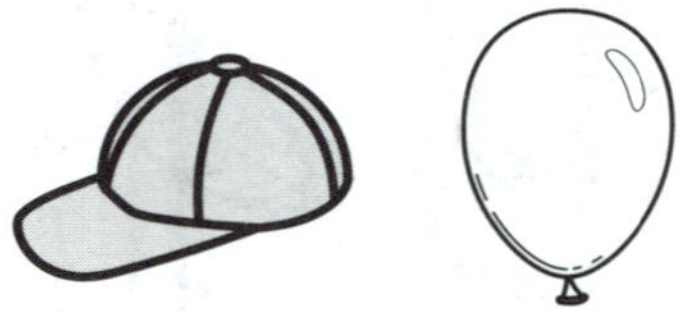

Chapter 10

✂ cut on all dashed lines   📄 fold on all solid lines

Which is above?

above

Which is below?

below

Chapter 10 Visual Kinesthetic Learning   VKV23

encima

debajo

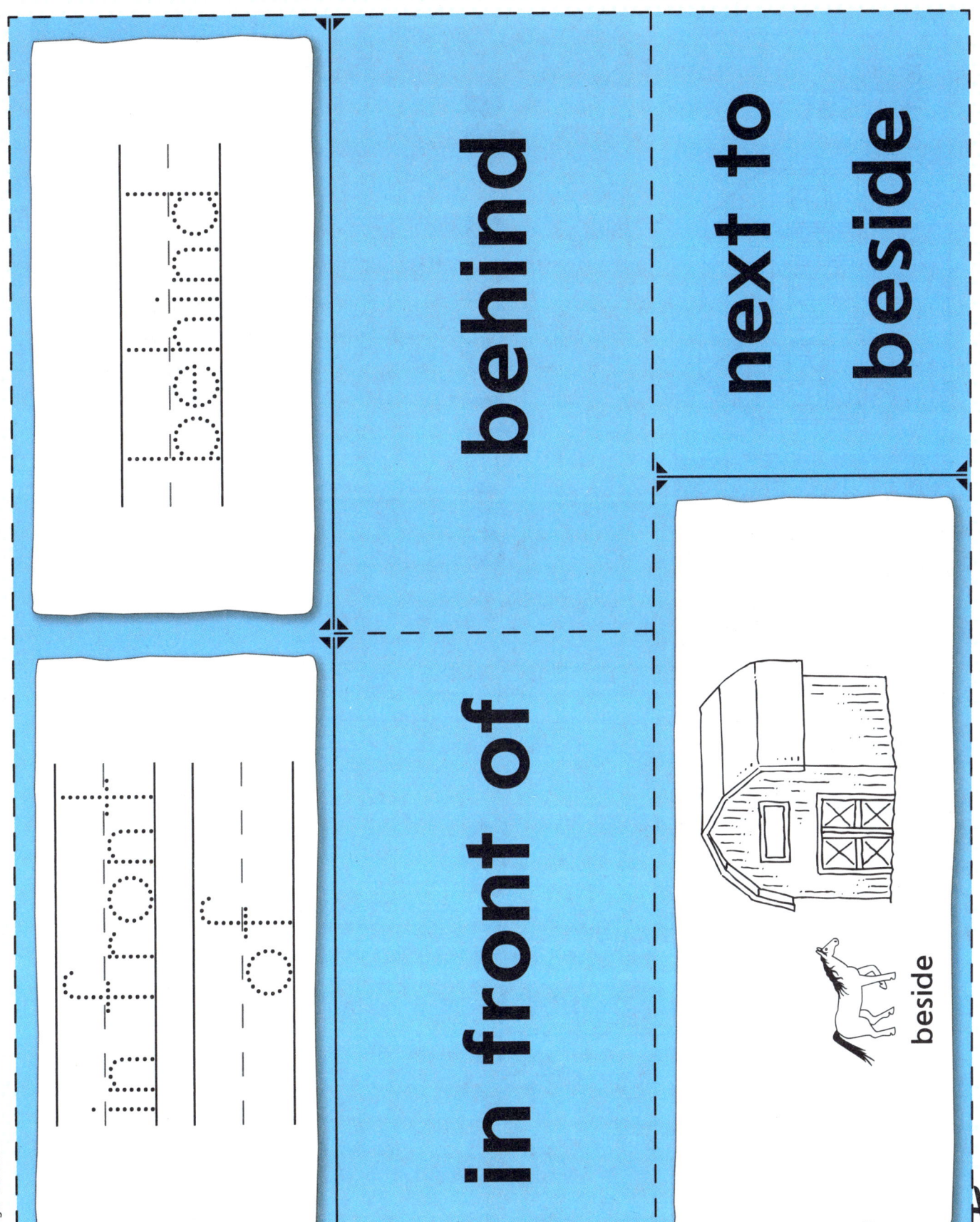

Dinah Zike's Visual Kinesthetic Vocabulary®

Chapter 10

cut on all dashed lines   fold on all solid lines

next to

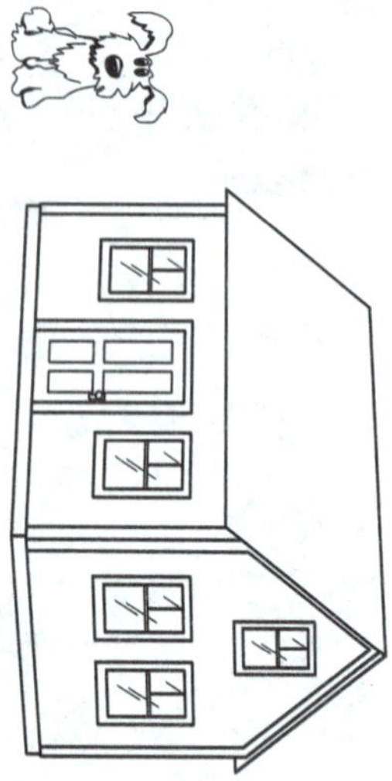

en frente de

junto a   al lado

detrás

VKV26   Chapter 10 Visual Kinesthetic Learning

Chapter 11

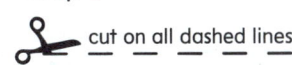

cut on all dashed lines   fold on all solid lines

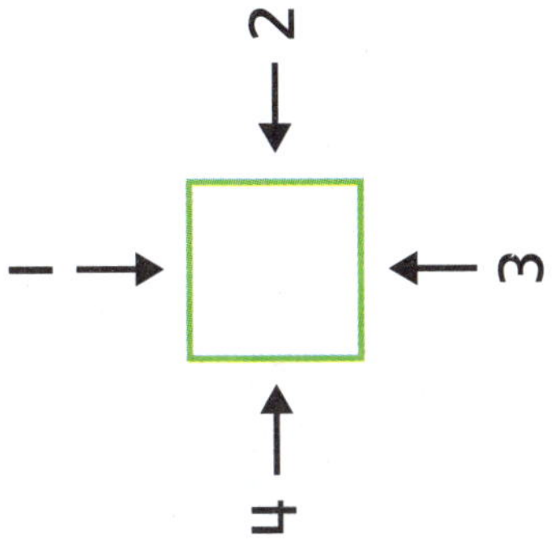

Which has a vertex?

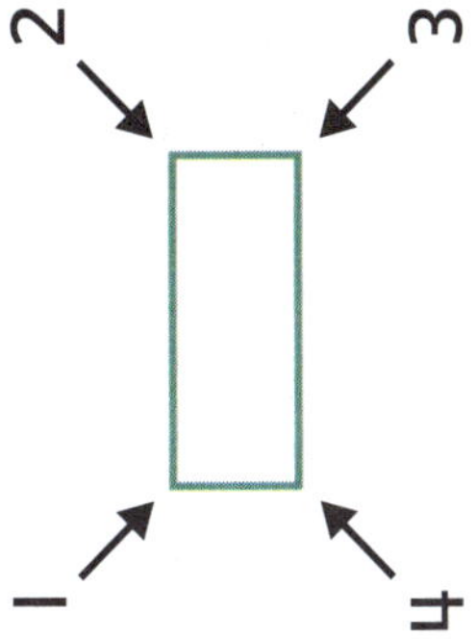

square

rectangle

vertex

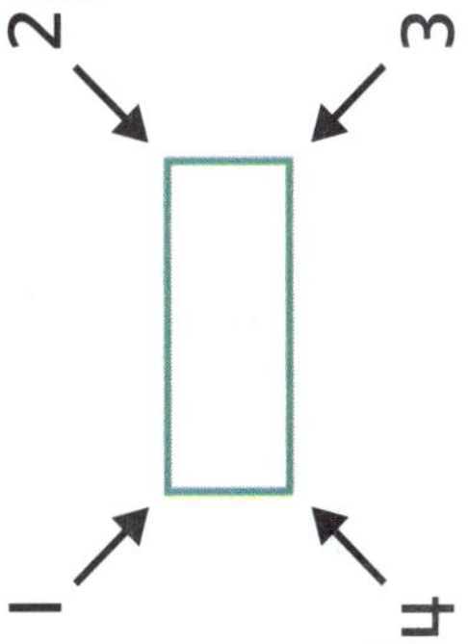

Chapter 11 Visual Kinesthetic Learning VKV27

Chapter 11

ice

ángulo

cuadrado

vertex

vé

Which is a rectangle?

Which is a square?

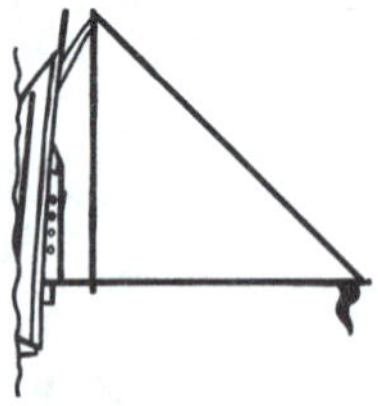

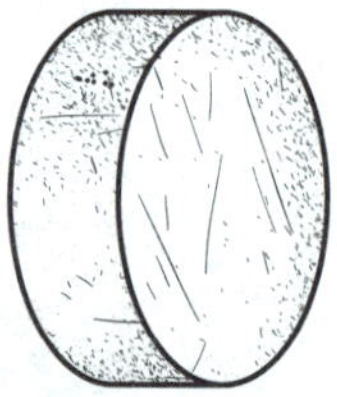

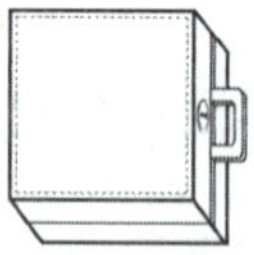

VKV28 Chapter 11 Visual Kinesthetic Learning

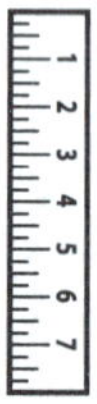

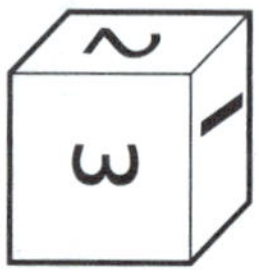

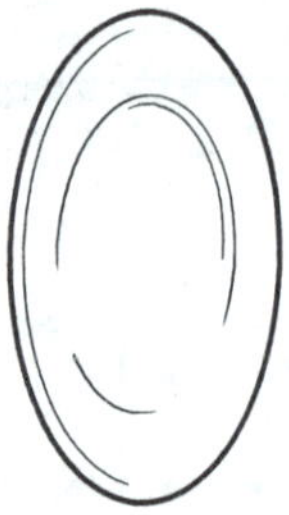

Chapter 11

cut on all dashed lines   fold on all solid lines

circle

___ vertices   ___ sides

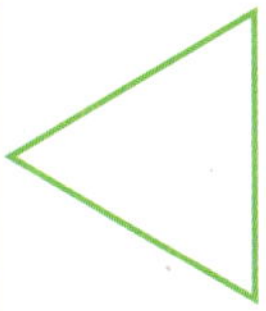

triangle

___ vertices   ___ sides

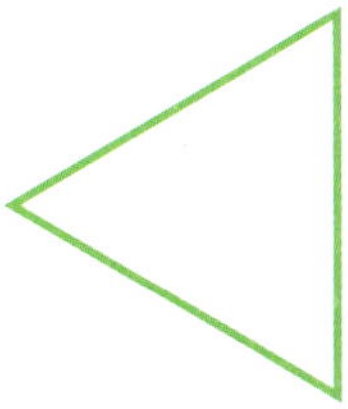

hexagon

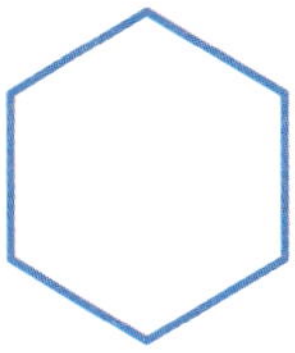

___ vertices   ___ sides

Chapter 11 Visual Kinesthetic Learning   VKV29

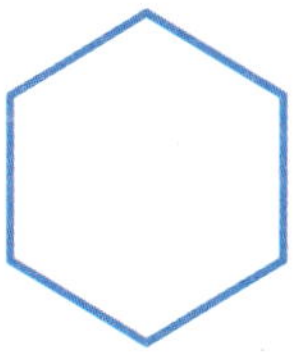

 Dinah Zike's Visual Kinesthetic Vocabulary®

Chapter 11

✂ cut on all dashed lines 📄 fold on all solid lines

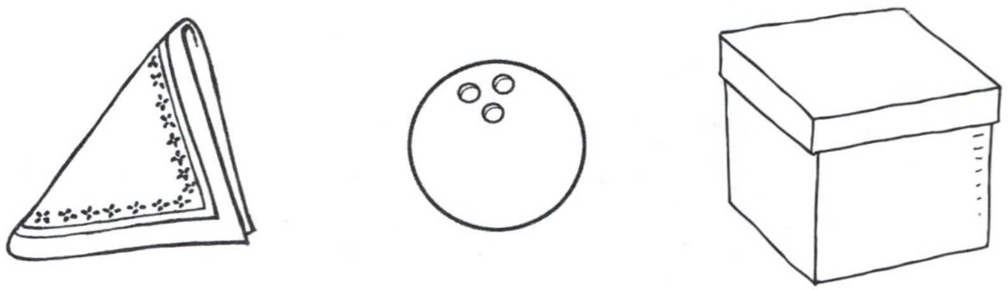

ágono

ángulo

írculo

Which is a hexagon?

Which is a triangle?

Which is a circle?

VKV30 Chapter 11 Visual Kinesthetic Learning

Chapter 12

cut on all dashed lines   fold on all solid lines

Which is a sphere?

Which is a cube?

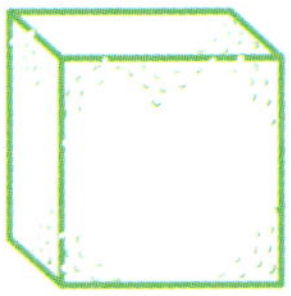

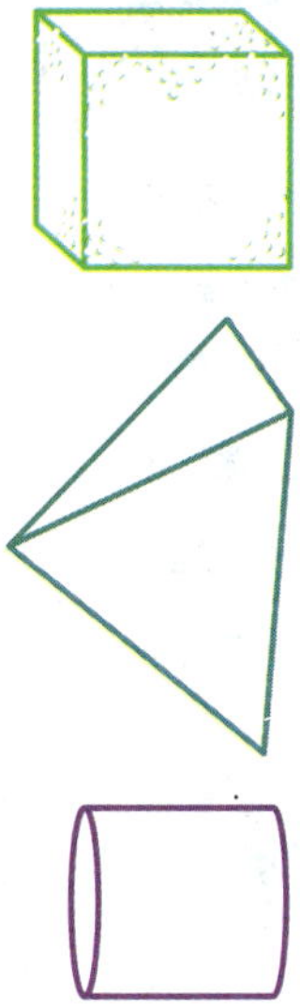

Which is a cylinder?

sphere

cube

cylinder

Chapter 12 Visual Kinesthetic Learning   VKV31

Chapter 12

✂ cut on all dashed lines   ▭ fold on all solid lines

ilindro   o   esfera

cylinder   cube   sphere

VKV32   Chapter 12 Visual Kinesthetic Learning

Chapter 12

Chapter 12 Visual Kinesthetic Learning VKV33

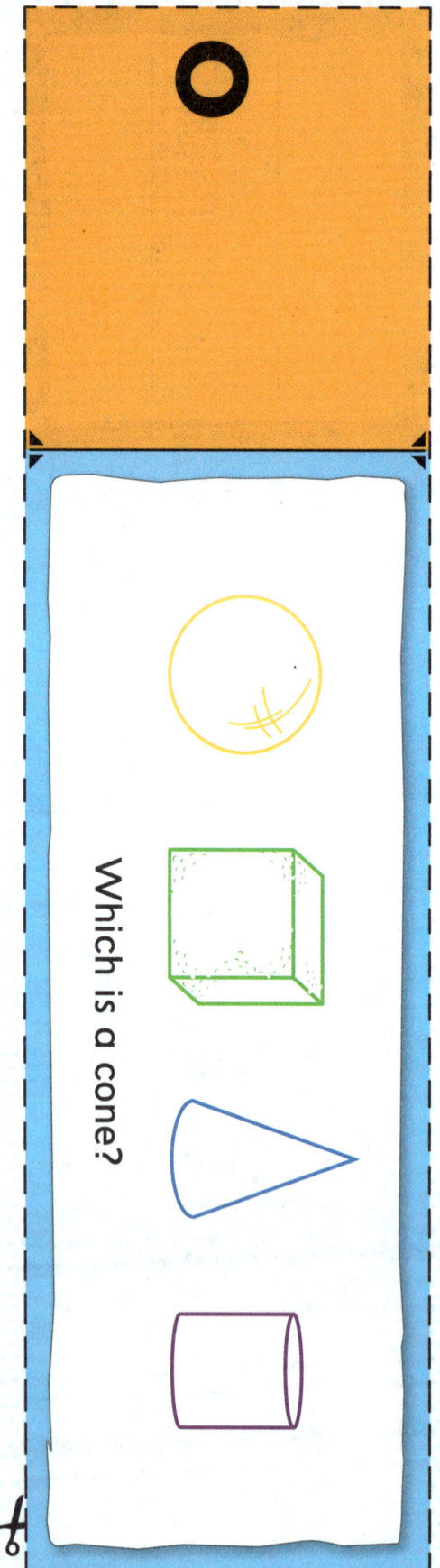